GREEN DEVELOPMENT

The 'green' debate has been gaining ground in development circles over the last decade. Environmentalists have been watching with interest. In *Green Development*, Bill Adams attempts to bridge the gulf that exists between environmentalism and development. He argues that development and environmental management should be seen as political processes.

The book traces the evolution of the concept of 'sustainable development', a phrase popular with environmentalists and development theorists alike. It analyses the ideas in *World Conservation Strategy* and the Brundtland Report and discusses radical 'green' development ideas. It also examines environmental aspects of development projects and programmes in tropical environments. It discusses the potential for 'green' reform of development planning, and in particular the technique of environmental impact assessment and its effectiveness in practice. It demonstrates the links between the impacts of development on the environment and those on people, particularly the rural poor.

Green Development questions the established understanding of the problems of environment and development, highlighting the inadequacy of a narrow view of environmental impacts, and a narrow response based on traditional conservation measures. It argues that the central focus of 'green' development has to be the environmental needs of the poor. Sustainable development planning must therefore focus on the power of the poor to control the environment on which they depend, and their capacity for self-determination in the face of development.

Bill Adams is a lecturer in Geography at the University of Cambridge. He has written *Nature's Place: conservation sites and countryside change* (1986), and has co-edited *Irrigation in Tropical Africa: problems and problem-solving* with A. T. Grove (1984) and *Environmental Issues in African Development Planning* with J. Seeley (1988).

Routledge Natural Environment — Problems and Management Series
Edited by Chris Park,
Department of Geography, University of Lancaster

The Roots of Modern Environmentalism
David Pepper

The Permafrost Environment
Stuart A. Harris

Environmental Policies: An International Review
Chris C. Park

The Conservation of Ecosystems and Species
G.E. Jones

Environmental Management and Development in Drylands
Peter Beaumont

Chernobyl: The Long Shadow
Chris C. Park

Nuclear Decommissioning and Society: Public Links to a New Technology
Edited by Martin J. Pasqualetti

Radioactive Waste: Politics and Technology
Frans Berkhout

Environmental Policy and Impact Assessment in Japan
Brendan F.D. Barrett and Riki Therivel

The Diversion of Land: Conservation in a Period of Farming Contraction
Clive Potter, Paul Burnham, Angela Edwards, Ruth Gasson and Bryn Green

GREEN DEVELOPMENT

ENVIRONMENT AND SUSTAINABILITY IN THE THIRD WORLD

W. M. ADAMS

London and New York

First published 1990 by Routledge
11 New Fetter Lane, London EC4P 4EE

Simultaneously published in the USA and Canada
by Routledge
a division of Routledge, Chapman and Hall, Inc.
29 West 35th Street, New York, NY 10001

Reprinted 1991 (twice)

Printed and bound in Great Britain by
Antony Rowe Ltd, Chippenham, Wiltshire

British Library Cataloguing in Publication Data

Adams, W.M. (William Mark), *1955–*
Green development: environment and sustainability in the Third World. – (Routledge natural environment problems and management series).
I. Developing countries. Economic development. Environmental aspects
I. Title
330.91724

ISBN 0-415-00443-8

Library of Congress Cataloging in Publication Data
applied for

This book is dedicated to my father,
who died while it was being written

H.C. Adams
1917–1987

Contents

List of Tables

Preface

This book was conceived in 1984, and largely written in 1987 and 1988. In the intervening period a great many changes took place in the ways in which those active in development viewed the environment, and the ways in which environmentalists understood development. There has been an astonishing growth in supposedly 'green' ideas and statements of intent from development agencies, up to and including the World Bank. This remarkable phenomenon has been enthusiastically welcomed, and indeed fostered, by environmentalists.

Partly for this reason, the scope of the book has shifted through time. There is now much more sophistication in debates about environment and development than there was even five years ago, as well as a much larger and more informed literature. My own understanding of the issues has also changed a great deal, partly as I have myself become aware of the scope of radical critiques of development and the relations between people, societies and nature. I have moved over the last ten years from a rather uncritical environmentalist position on development, chiefly informed by ecology and chiefly dominated by issues of conservation, to a more complex position, and I hope a richer understanding of the issues. In a sense I have had the pleasure, and the pain, of exploring the context and implications of my own beliefs.

This book has therefore been written while rising on a steep learning curve, moreover one that still remains steep. It has been exhilarating to write, although the resulting book probably still reveals a certain excess of enthusiasm and naïvety. In retrospect, there are also some notable omissions. I regret writing so little about the essentially global concerns of the greenhouse effect, ozone and transboundary air pollution. These have risen in importance, and on the international and environmental agendas, very rapidly in the late 1980s. I am also conscious that the book is anglocentric in its coverage of literature, and excessively biased towards Africa. For this the only mitigation is the importance of writing from experience wherever possible. Most of my work in the Third World has been in Africa, and as a result I am at least better aware in that context of the scope of my ignorance. Where experience elsewhere differs, I hope others are impelled to describe it.

The book is an attempt to bridge two important gulfs. The first is between environmentalism and development, and the second between armchair theorists and practitioners. To create the first bridge I have devoted most of the first four chapters of the book to a discussion of global issues, particularly of those theoretical ideas that have gone to form thinking about sustainable development, and the ways these relate to debates in development. These chapters argue that environmentalists still have much to learn from radical views of political economy, although (regrettably, but perhaps sensibly) I do not pretend that they offer a simple shortcut to an adequate radical 'green' theory.

The second gulf which the book tries to bridge is that between the ivory tower (in which it will be clear that this has been written) and the real world of development decisions. To tackle this, later chapters explore environmental problems at smaller scales, focusing on environmental aspects of development projects and programmes. The environmentalist challenge to development has been mounted on the basis of the critique of development practice, and calls for the reform of environmental planning procedures. Therefore, I have discussed both the nature of the environmental impacts of development, and the very limited success (and limited prospects) of reformist measures such as Environmental Impact Assessment in preventing them.

I have explored the problems that face those environmentalists who expect 'sustainable development' to serve effectively as a means to further traditional nature conservation concerns. Development can certainly be coercive, as well as environmentally damaging, but so too can conservation. Sustainable development is a beguiling slogan, but as a front for a concern about wildlife alone it is potentially a cynical and empty one. The radical environmentalist critique is deeper (and far less convenient) than it might at first appear.

Ultimately this book argues that the 'greening' of development calls for a quite fundamental reassessment of the question of development itself. The environmentalist critique of development which sees reform as basically needing relatively minor changes of planning procedure and better protection of key areas for conservation is, in my view, quite inadequate. However, this kind of thinking dominates the process of 'greening' that has taken place. It is the focus of documents on sustainable development such as the *World*

Conservation Strategy and the Brundtland Report, and the main feature of the arguments of environmentalist pressure groups. It has been the key element also in the proposals of the major aid agencies to reform their environmental procedures. However, behind this pragmatic (and within limits effective) approach lies a deeper and more subversive vision concerned with the nature and the scale of power over environment and people. The word 'development' implies that this power is held by states and their international advisers. Reforming the way that power is used does nothing to alter the capacity of people to plan and run their own lives, or control their own environment. Rather than contributing to the enhanced capacity for centralized bureaucratic and technocratic power, 'green' development could foster the capacity of individuals and groups to control their own resources.

Having started with ideas about sustainable development, the book moves on to a discussion of the practice of environmental concern in development, and ends with this notion of the need to place power back with the powerless in the rural Third World. This is, of course, a notorious Pandora's Box. Here it is put forward with some enthusiasm, with perhaps also too much naïvety. Furthermore, while the box is opened, by no means all the questions that swarm out, buzzing angrily, are tackled adequately. Perhaps this is unreasonable, but to attempt that task would require another book.

If there is a single conclusion to be drawn from what I have written, it is simply this: it is no good talking about how the environment is developed, or managed, unless this is seen as a political process. Any understanding of environment and development must come back to this issue. The 'greenness' of development planning, therefore, is to be found not in its concern with ecology or environment *per se*, but in its concern with control, power and self-determination. Will environmentalists adopt the rather subversive view of development advocated here? Should they do so? I ask these questions and am convinced of their importance. I am only too well aware that this book only begins the task of answering them.

Acknowledgements

A number of people have helped in the writing of this book, some of them unwittingly. I am grateful to Chris Park for suggesting that I write it, and for accepting it when I had done so. David Lea, Stuart Corbridge, Dick Grove, Andrew Warren, Ted Scudder and Cathy Nesmith were kind enough to read all or parts of the manuscript. Those errors which remain do so despite their efforts. I am also grateful to those (both experts and fellow-sufferers) who have helped guide me through the three word-processing systems in which the text has been buried at different times.

While researching the book, I have learned a great deal from Martin Adams, and those who contributed to the IIED/Danida workshop on environmental strategy in semi-arid areas in Copenhagen in 1988. I am grateful also to Dave Anderson and Richard Grove, who organized a workshop on conservation and development in Africa in Cambridge in 1985, at a time when I had just begun to work on this book, and for all those people who helped organize and participated in seminars on environment and development organized through the African Studies Centre from 1985 to 1988. Most of them will probably think I have learned little from our deliberations, but that would be far from the truth. I also owe a considerable debt to those I have worked with on development projects in Africa, for enabling me to enjoy the pleasures and pains of learning by doing, and showing forbearance when I wanted to argue about it.

Above all I would like to thank Francine for all her support. She has read the book (in bits and together) several times, and commented extensively. She has shown both a high pain threshold and a sympathy beyond all that is reasonable. Emily and now Thomas have also helped in their ways, if only by making opportunity for meditation in the still watches of the night, and providing necessary and very personal deadlines.

1

Greening Development?

Environmentalism and development

Within the last decade, concern about environment and development in the Third World has become a central feature of the rhetoric and thinking of development studies. Of course, awareness of the environmental aspects of development is not by any means new, whether among scholars, practitioners or participants in development. What is new is the scope and sophistication of environmentalist critiques of development in practice, and the new focus by some development thinkers on the environment in the context of social and economic change.

To environmentalists, the loss of natural habitat and breakdown in indigenous and subsistence ways of life caused by development projects has been a potent focus for the extension of the pressure group politics familiar in the industrialized world since the 1970s. 'Save the rainforest' campaigns, or the articulation of opposition to investment in large projects like dams, follow logically enough from concerns about pollution, whales or First World rural environments. To an extent, concern for Third World environments simply reflects the growing integration of the global village, and environmentalist pressure is at least in part an extension of traditional concerns about environmental quality in that village's new countryside, in the Third World. First World environmentalism, however, has done more than simply broaden its global scope. There has been a self-conscious effort to move beyond environmental protection and transform conservation thinking by appropriating ideas and concepts from the field of development.

The evidence of this transformation in environmentalist rhetoric is legion, but much of it focuses on the almost universal adoption of the phrase 'sustainable development'. The concept of sustainable development was important at the 1972 United Nations Conference on the Human Environment in Stockholm. Subsequently it was taken up by a number of authors (e.g. Riddell 1981, Sachs 1979, 1980, Glaeser 1984a, b). It is variously defined: Eckholm (1982), for example, sees it in terms of the failure of the usual 'pell-mell, lopsided development'. He calls instead for 'economic progress that is ecologically sustainable and satisfies the essential needs of the underclass' (Eckholm 1982, p. 8). International agencies such as the United Nations Environment Programme (UNEP) and the International Union for Conservation of Nature and Natural Resources (IUCN) have also backed sustainable development. It was the central concept in the World Conservation Strategy published in 1980 (IUCN 1980) and is also the foundation of the report of the World Commission on Environment and Development seven years later (Brundtland 1987). These two documents underpin most contemporary environmentalist thinking on sustainable development.

However, sustainable development has gained a currency well beyond the confines of global environmental organizations. It has become, in Conroy's term, the 'new jargon phrase in the development business' (Conroy 1988, p. xi). It has also been the subject of a series of studies (e.g. Redclift 1984, 1987, Douglass 1984, Clark and Munn 1987, Carim *et al.* 1987), and has been taken seriously by some economists (Goodland and Ledec 1984, Pearce 1988a, 1988b, Turner 1988a).

Sustainable development received the imprimatur of the international development establishment in 1987. In January of that year the British Minister for Overseas Development spoke to the Royal Geographical Society in London on Aid and the Environment, stressing the importance of environmental issues in development planning (Patten 1987). In the following April the World Commission on Environment and Development published *Our Common Future* (Brundtland 1987), setting out a 'global agenda for change' which was centred on the concept of sustainable development. In the same month the International Institute for Environment and Development (IIED) held a conference in London (Conroy and Litvinoff 1988), and in May the President of the World Bank spoke of the links between ecology and sound economics in a major statement of the Bank's policy on the environment (Hopper 1988).

This display of conservative development agencies and environmental groups and organizations dancing to the same tune is remarkable, but not entirely accidental. It reflects in part the success of environmentalist pressure on aid donors, backed by willing and effective media coverage (e.g. Timberlake 1987), combined with renewed concern about the global environment, particularly the 'ozone hole' and the 'greenhouse effect', with an international conference on ozone in London, and an 'environmental summit' in the Netherlands in

March 1988. In part the response of development agencies to environmental issues also reflects what may be a more general 'greening' of politics in Western industrial countries (see Capra and Spretnak 1984, Rudig and Lowe 1986, 1989, Hay and Hayward 1988). The rhetoric of development has certainly been affected by this wave of relabelling, and the 'greening' of development thinking has been celebrated (e.g. Harrison 1987, Conroy and Litvinoff 1988).

The question remains, however, how deep this apparent revolution goes. Has there really been a 'greening' of development? Has there, for example, been a revolution in ideology in any way analogous to Charles Reich's celebrated account of new thinking in America in the 1960s (Reich 1970)? At a time when there is such visible enthusiasm about new perspectives and new alliances over the environment, this is a hard and unpopular question, but it is an important one. The answer turns on the extent to which 'sustainable development' or 'green' development or 'ecodevelopment' are words backed up by logical theoretical concepts rather than simply convenient rhetorical flags under which ships of very different kinds can sail.

Unfortunately, sustainable development thinking very often proves to have no coherent theoretical core. With some reason, Redclift comments that sustainable development 'seems assured of a place in the litany of development truisms' (Redclift 1987, p. 3). It is a phrase with many meanings. The commonest definition now is that of *Our Common Future*, 'development which meets the needs of the present without compromising the ability of future generations to meet their own needs' (Brundtland 1987, p. 43). As a definition, this is superficially attractive, but it is a better slogan than it is a basis for theory. Commentators strew phrases like sustainable development and related terms like 'ecodevelopment' around with great freedom, but seldom give them a clear and consistent meaning.

Indeed, such phrases have a terrible versatility, allowing users to make high-sounding statements with very little meaning at all. The flexibility of these terms, and their 'beguiling simplicity' (O'Riordan 1988, p. 29) of course adds greatly to their attraction. On the one hand the label is used by environmentalists to try to demonstrate the relevance of their ideas about proper management of natural ecosystems. The conviction behind works such as the *World Conservation Strategy* is that sustainable development is a concept which truly integrates environmental issues into development planning. In using terminology of this sort, environmentalists have attempted to capture some of the vision and rhetoric of development debates. Sadly they often have no understanding of their context or complexity. Environmentalist prescriptions for development, shorn of any explicit treatment of political economy, can have a disturbing naïvety.

On the other hand the phrase 'sustainable development' is attractive to development agencies and theorists looking for new labels for liberal and participatory approaches to development planning. Development bureaucrats and politicians have undoubtedly welcomed the opportunity to fasten onto a phrase that suggests radical reform without actually either specifying what needs to change or requiring specific action. Thus the UK response to the Brundtland Report emphasizes the continuity between 'strands which have previously had separate currency' in British government policy and the concept of sustainable development, or 'development without destruction' which is presented by the report (Department of the Environment 1988). As is often the case, sustainable development is apparently acceptable to the British government precisely because it does not demand radical change of policy direction. As O'Riordan acidly comments, 'developers now realize that under the guise of sustainability almost any environmentally sensitive programme can be established' (O'Riordan 1988, p. 31).

One reason for the overlapping meaning of sustainable development is the highly confused question of what development itself means. This is a semantic, political and indeed moral minefield (Goulet 1972). Seers (1977) argues that if poverty, inequality and unemployment are decreasing without a loss of self-reliance (for example through foreign ownership of manufacturing plants), then development is taking place. Versions of this formula are widely debated, and are elaborated in the many textbooks on development (e.g. Mabogunje 1980, Forbes 1984), but development itself nonetheless remains an ambiguous and elusive concept, prey to prejudice and preconception. It is 'a Trojan Horse of a word' (Frank 1987, p. 231), a term which is sufficiently empty that it can be filled at will by different users to hold their own meanings and intentions. Concepts of development have a complex pedigree and etymology. There is a long history of 'development thinking before development began' (Brookfield, 1975, p. 2) dating back to the rise of mercantilism and economic liberalism which slowly - largely through the emerging colonial empires of the later nineteenth century - gave way to the emergence of a specific ethical ideology of 'development'.

That ideology presents 'development' as the process that recreates the industrial world: industrialized, urbanized, democratic and capitalist. It is often depicted as a 'crucible' through which successful societies emerge purified, both modern and affluent (Goulet 1971). This capitalist and eurocentric 'developmentalism' (Aseniero 1985) has come to define and dominate development thinking both within and outside the countries of the 'Third World' (or its equally euphemistic and disparaging alternative labels, the 'South', 'developing' or 'underdeveloped' countries).

The developmentalist ideology suggested that countries developed through different stages, on 'a linear path towards modernisation' (Chilcote 1984, p. 10), and progress down that path could be measured in terms of the growth of the economy, or some economic abstraction such as per capita gross domestic product. 'Development' meant the projects and policies, the infrastructure, flows of capital and transfers of technology which were supposed to make that imitation possible. It was the imposition of the established world order on the newly independent periphery. Illich comments, 'There is a normal course for those who make development policies, whether they live in North or South America, in Russia or Israel. It is to define development and set its goals in ways with which they are familiar, which they are accustomed to use in order to satisfy their own needs, and which permit them to work through the institutions over which they have power or control'. He concludes: 'this formula has failed and must fail' (Illich 1973, p. 368).

However, as the United Nations First Development Decade (1960-70) wore on, the certainties of developmentalism faltered. Frobel *et al.* argue that social and economic conditions for the majority of the population in many of the countries of the capitalist periphery 'steadily worsened in the course of the immediate post-war years' (Frobel *et al.* 1985, p. 114). Certainly commentators from a wide range of persuasions began to admit (and theorize about) the glaring gap between bland and simplistic expectation and reality. Debate about the nature and causes of the apparent failure to 'develop' has created the burgeoning disputes of development studies, and the proliferation of development theory (Chilcote 1984, Forbes 1984, and Corbridge 1986).

The failings of orthodox approaches to development, however, have also increasingly fuelled criticism of a different order from environmentalists. Ideas like sustainable development have picked up some elements of critiques of the development process, drawing rather indiscriminately from populist writings (for example about failures of distribution and the plight of the poorest), from radical ideas (such as dependency theory), and from more pragmatic critiques of development project appraisal and implementation. To this have been added concerns about the environmental impacts of development, both the costs in terms of lost ecosystems and species, and (latterly) the impacts of development action on natural resources for human use. Above all, environmentalist critiques of development have presented a picture of the Third World in crisis. The notion of global crisis was an important element of environmentalism in the 1960s and 1970s. It has become a central element in debates about sustainable development. In this it is, of course, matched by concerns within development itself.

Crisis in the Third World

Crisis has become a commonplace motif of development writing (e.g. Brandt 1983, Timberlake 1985, Ravenhill 1986, Lawrence 1986, Anderson and Grove 1987). The perception of dramatic and unsolvable problems in the countries of the 'South' is common to politicians, aid agencies, academic analysts and the media. The 1984 drought in Sahelian Africa intensified this perception, leading of course to astonishing individual generosity of those unkindly (but perhaps accurately) described as 'the fat and happy in rich countries' (Harrell-Bond 1985, p 13). This indicates the disturbingly durable nature of the resulting picture of poverty: paradoxically it seems possible for the Third World to be perceived as being locked in a permanent state of crisis of politics, economics and environmental degradation.

This view of crisis tends to favour 'firefighting' approaches to development as against discussions of deeper ills, and the treatment of symptoms not causes. Thus Julius Nyerere commented in 1985 'African starvation is topical, but the relations between rich and poor countries which underlie Africa's vulnerability to natural disasters have been relegated to the sidelines of world discussion' (Nyerere 1985). This urgency leaves little time for lateral thinking. The magnitude of problems and the failure of conventional economic thought and the bureaucracies to solve them has made sustainable development widely attractive in the 1980s.

Literature on the Third World suggests that there are two distinct crises, of development and of environment. The crisis of development, or the lack of it, is perhaps the better studied (see for example Frank 1981, Frobel *et al.* 1985, Edwards 1985). It embraces the problems of debt, falling commodity prices, falling per capita food production, growing poverty and socio-economic differentials both within Third World states and between countries. The dimensions of the problems facing the Third World are substantial and real enough. The litany of statistics is grimly familiar. Life expectancy at birth in low income countries in 1982 was 52 years for women and 50 for men, compared to 78 and 71 respectively in the Western industrial market economies (World Bank 1984b); average population per doctor was 15,000 in low-income countries in 1980 compared to 554 in the industrial North. The death rate of children aged between 1 and 4 in 1982 was 19 per thousand in low income countries, reaching over 3 per cent in Chad, Burkina Faso, Guinea, Sierra Leone and Afghanistan (World Bank 1984b). Obviously the accuracy and usefulness of aggregate statistics of this sort is limited. Nonetheless, the overall picture, of the essentially human dimension of the Third World's crisis, is clear.

Analyses of the cause of the problems behind this crisis of development vary. On one side there is a great diversity of radical commentators, for example Frobel *et al.* (1985), who identify a crisis of 'the world capitalist system as a whole, including much of the Third World too' (Frobel *et al.* 1985, p. 111), and a 'collapse in the credibility of the capitalist economic and social system' (Frobel *et al.* 1985 p. 112). On the other side are the world financial institutions, spearheaded by the implacable economists of the World Bank. In 1984 they commented that the recession was over, having lasted from 1980 to 1983 the 'longest in fifty years'. However, it provided 'many valuable lessons for economic policy because it highlighted longstanding weaknesses in every economy and in international arrangements', and had lessons to teach if recovery was to mature into 'sustained and rapid growth of the kind the world enjoyed for twenty-five years after World War Two' (World Bank 1984b, p. 1).

Unlike the previous shorter recession in 1974-5, the amount of foreign capital (in the form of aid or commercial loans) available to Third World countries declined rapidly after 1981, while the indebtedness rose. The external public debt of low income Third World countries (excluding China and India) rose from about 21 per cent of GNP in 1970 to over 28 per cent in 1982, with no less than seven countries having a debt of over 50 per cent of GNP (Mauritania an unbelievable 146 per cent, Togo 104 per cent, Mali 79 per cent, Zaire and Somalia 78 per cent). The situation in middle-income countries was if anything worse, with a significant rise in average indebtedness and debt larger than GNP in Mauritania, Costa Rica, and Nicaragua.

Set against the grim picture of intensifying debt in the early 1980s was the failure of agriculture to keep pace with foodgrain production. Sub-Saharan Africa was particularly hard hit, with intense droughts in the Sahel and elsewhere in the continent in the early 1970s and the early 1980s, agricultural policies that discriminated against the small farmer in favour of costly, intensive, large-scale projects and the irrelevance (and hence ineffectiveness) of 'green revolution' technology. The evidence suggests that per capita food production fell in sub-Saharan Africa consistently through the 1960s and 1970s.

The second crisis of the third World portrayed in the literature is environmental. It embraces desertification and fuelwood shortage (Grainger 1982, Munslow *et al* 1988), and the logging of tropical rainforest (Routley and Routley 1980, Caufield 1982, Myers 1984). Some recognize a third crisis, of energy, separating out fuel and particularly fuelwood problems. These problems and others are widely reviewed, for example by Myers (1985) and the World Resources Institute (1986), and have become an important focus for environmental concern in the 1980s. From 1972 the United Nations Environment Programme (UNEP) published annual 'state of the environment reports' on particular issues, and carried out a review of 'world environmental trends', ranging from atmospheric carbon dioxide and desertification to the quality of drinking water (Holdgate *et al.* 1982a, b).

Environment and development

The danger of global crisis in development, environment and energy is a major theme of the Brundtland report. However, in practice, the fields of developmental and environmental studies are far from unified, often being remote from each other both conceptually and practically. So-called 'experts' (in the sense used by Chambers 1983) rarely even claim expertise in both, and seldom understand linkages. Sociologically, the two fields have their own separate cadres and culture, their own self-contained arenas of education and theory formation, their own technical language, research agendas and - above all - their own literature. Each represents a distinct culture. Although the two cultures obviously overlap a great deal, and indeed make confident inroads onto each other's territory with scant regard for the exact meaning or purpose of terminology, there is rarely if ever any integration.

It is indeed only rarely and recently that environment and development have been linked theoretically with any kind of success. This has now been done by arguing the need to set environmental resources and resource use in a social and political as well as an economic context. Radical political economy in particular offers a powerful way of doing this (see Blaikie 1985, Blaikie and Brookfield 1987, Redclift 1984, 1987). Why has this linkage been so elusive for so long? One reason is the mechanistic thinking and chronic tunnel vision of economists. This is backed up by parochial and doctrinaire thinking by radical theorists, constrained by 'oppositionism' and 'determinism' (Corbridge 1986). Behind this again lies a more general failure by both academics and practitioners to cross disciplinary boundaries.

The notion of 'disciplinary bias' as a constraint on innovative thinking about complex development fields is firmly established (Chambers 1983), and it is clear that there is an unavoidable and hence critical ideological component in understanding problems of environmental resource use and environmental degradation. The best worked-out example of the central importance of ideology in understanding such 'environmental' phenomena is that of soil erosion and erosion control projects planned and enforced by the state, particularly in colonial territories (see Blaikie 1985, Blaikie and Brookfield 1987, Anderson 1984). This issue is discussed in depth in

Chapter 5. It is a persistent failure of environmental scientists, and of environmentalism, to refuse to recognize the ideological burden of ideas and policies. People trained in natural science disciplines in particular find it difficult to transcend the notion of the impartiality and 'truth' of science, and hence to agree a common approach to understanding - let alone tackling - field problems of poverty and environment (Seeley and Adams 1987). These problems of disciplinary biases are discussed in Chapters 7 and 8.

Development crises and environmental crises exist side by side in the literature, and together on the ground, yet explanations often fail to intersect. Environmentalists and social scientists tend to speak different languages and there is little mutual understanding. Very often theirs is a dialogue of the deaf, carried on at cross purposes and frequently at high volume. There are close and reciprocal links between development, poverty and environment. The complex and multi-disciplinary nature of these links makes them difficult to identify and define. They often go unnoticed, fall down the cracks between disciplines, or get ignored because they fit so awkwardly into the structures of academic analysis or discourse. Nonetheless, in the real world these links are real enough. They explain why development policy often causes rather than cures environmental problems. Development and environmental degradation often form a deadly trap for the poor.

Chambers (1983) argues that it is the plight of the poor that should set the agenda for development action, and in his approach to sustainability he directs attention to the concept of sustainable rural livelihoods, defined as the secure access to sufficient stocks and flows of food and cash to meet basic needs. He suggests that there are both moral and practical imperatives for making sustainable livelihood security the focus for development action. This is the principle underlying the arguments in this book: the touchstone for debate about environment and development is the human needs of the poor, both environmental and more conventionally developmental. Among other things, it makes debate about the environment in the Third World, like that about development, inherently political; but, as Redclift argues, it is an illusion to believe that environmental objectives are 'other than political, or other than distributive' (Redclift 1984, p. 130).

Outline of the book

This book is not another attempt to find the winning formula, the mix of sticks and carrots, rhetoric, capital flows and environmental knowledge that will achieve 'real' development. Rather, it is about the peculiar difficulty of talking sensibly about the environmental dimension to development in the Third World. Its aims are, first, to discuss the nature and extent of the 'greening' of development theory. It does this by examining the key concept of sustainable development. It looks at the origins and evolution of these ideas, and offers a critique of their articulation in the World Conservation Strategy and the Brundtland Report. It is argued that the ideology of sustainable development is eclectic and often confused. Sustainable development is essentially reformist, calling for a modification of development practice, and owes little to radical ideas, whether claiming a green or Marxist heritage. Second, the book attempts to draw a link between theory and practice by discussing the nature of the environmental degradation and the impacts of development. In doing so it attempts to address the question of the limitations of reformist approaches. It argues that, ultimately, 'green' development has to be about political economy, about the distribution of power, and not about environmental quality.

The first part of the book is largely theoretical, and focuses in particular on the global scale. Chapter 2 discusses the origins and growth of sustainable development ideas, looking in particular at their roots in nature preservation, colonial science, and the internationalization of scientific concerns in the 1960s and 1970s. This is an account of institutions and organizations as well as ideas, and takes the form of a historical account. Attention is focused on the 1970s and the Conference on the Human Environment in Stockholm in 1972, at which sustainable development became a specific and identified area of concern. The World Conservation Strategy was a direct development of the thinking at that time.

Chapter 3 discusses the ideas in the World Conservation Strategy and the Brundtland Report in detail, analysing both what they say, and the nature of their ideologies. The realism of sustainable development ideologies is discussed, and their capacity to theorize about society, economy and the role of the state and the world economy is examined. Other ideas of sustainable development or 'ecodevelopment' are also considered, and are put into the context of conventional development thinking. Most of these ideas are best seen as versions of neo-populism, concerned at the reform and not the transformation of the development process. However, there are more radical ideas about world development, and these are discussed in Chapter 4. The arguments of radical political economy are briefly reviewed, and the related and often derived ideas of radical green thinkers are analysed. These ideas are often sharply divergent from both the conventional reformism of sustainable development thinking and the mainstream of radical thought.

The second half of the book provides a commentary on the theoretical ideas in the first half by discussing environment and development in practice. It moves down the scale continuum to focus attention on development projects. In Chapter 5 the links between development and environmental degradation are drawn out in the specific context of arid land desertification in Africa. The nature of arid land degradation and the difficulty of defining desertification are discussed. The nature of ideology in influencing attitudes to, and perception of, degradation is related to the political economy of desertification. The political and economic implications of policies developed to counteract desertification are analysed, and related to their sustainability.

In Chapter 6 the environmental impacts of developments are considered, in order to draw out the links between development action and the environment. It is in this area of environmental impacts that the environmentalist critique of development has been most effective, and this chapter expands the established environmentalist critique of development projects. Particular attention is paid to rainforest development, and the physical and ecological impacts of dam construction. Tropical wetlands are then used as an example of the integrated nature of environmental impacts. Chapter 7 extends this debate by examining the nature and effectiveness of the established techniques of environmental appraisal, particularly Environmental Impact Assessment. This is the main focus of attempts to reform development practice, and this chapter analyses the limitations of this reformist approach, and the reasons for the failure to make apparently adequate planning structures effective in reducing environmental impacts. It discusses problems of taking environmental impacts into account in the real world of project construction contracts and aid-giving, and identifies where these initiatives for reform might be improved and made effective.

Chapter 8 moves on beyond reformism to put people back into the picture. It discusses the relations between development agencies and those who participate in development projects, too often a case of the blind mistreating the dumb. This relationship is seen most clearly in projects that involve resettlement. Evacuees from reservoirs or other project areas often bear disproportionate shares of the costs of developments, and miss out on benefits. One result of such treatment is the opposition of those affected by development projects, resistance being in the form of violent or non-violent action Lastly, people are used to draw links between conservation and other aspects of sustainable development. It is argued that, very often, wildlife conservation appears as just another form of alien state-imposed development. This obviously suggests that the apparently close links often drawn between conservation and development may involve at least a measure of wishful thinking.

Sustainable development, like environmentalism, is a blend of technocentrist and ecocentrist worldviews (O'Riordan 1981, O'Riordan and Turner 1983). Technocentrist approaches are rationalist and technocratic, and underlie approaches to the environment involving management, regulation and 'rational utilization'. Ecocentrism (or biocentrism, see Worster 1985) is romantic and transcendentalist in tradition, embracing ideas of bioethics and utopianism. Both perspectives are represented in thinking about sustainable development, often in a confusing mix of essentially technocentrist measures (such as Environmental Impact Assessment) and ecocentrist ideas (for example 'populist' or Gandhian 'development from below').

Within the breadth of thinking about sustainability there is clearly some common ground. Turner (1988b) suggests that a coalition may be possible between 'accommodating technocentrism' (a conservationist position of sustainable growth) and 'communalist ecocentrism' (a preservationist position emphasizing macroenvironmental constraints on growth, and decentralization). Such a coalition would exclude the more radically opposing worldviews of 'cornucopian technocentrism' and 'deep ecology ecocentrism'. However the field is classified, there is no doubt that current visions of sustainable development are rather messy: enthusiastic, positive and committed, without, in general, being overtly political. Precise meanings remain elusive (Turner 1988b), and in real life the distinction between ecocentrism and technocentrism is blurred, and divergent values and ideas can (and do) occur together and overlay each other.

In the last chapter of this book, the twin strands of sustainable development, reformism and radicalism are drawn together. The result is not a synthesis, in the sense of a shopping list of environmental desiderata, nor an attempt to set out a blueprint for 'sustainable development'. There is no comfortable celebration of the achievements of the new green bankers and aid bureaucrats. Instead, what is revealed is a tension in the heart of the environmental critique of development practice which challenges the calculated reformism of sustainable development. Important elements in green critiques of development are radical and not reformist, and they are both awkward and inconvenient. O'Riordan (1988) distinguishes between the essentially technical concept of sustainable utilization and the broader, ethically-based, concept of sustainability. He argues that sustainable utilization is bureaucratically and politically acceptable precisely because of its ambiguity. This is the target of reformist thinking in environmentalist critiques of development. Sustainability, on the other hand, is 'politically treacherous', challenging to the status quo and radical. Between these two broad views there is tension. The ethics of sustainability demand rather more than merely reform of the development process. The 'greening' of development demands a more radical analysis, and a more transforming response.

In 1970 Charles Reich wrote with passion and hope in *The Greening of America* of the new consciousness abroad in America, 'arising from the wasteland of the corporate state like flowers pushing up through the concrete pavement' (Reich 1970, p. 328). Almost two decades later, the confident exuberance of that time is gone, but many of the seeds sown in the 1960s have taken root. Despite their flaws, the growth of sustainable development ideologies have had an impact on the consciousness that informs development thought and action. There is a new environmental awareness in development which is perhaps evidence of a 'greening' of a kind. However, to date it has been largely superficial, a thin green layer painted onto existing policies and programmes. There is plenty of 'green' talk from politicians and development bureaucrats, but limited evidence of radical change in ideas, aims or policies.

2

The Origins of Sustainable Development

Environmentalism and the emergence of sustainable development

The concept of sustainable development is at the centre of current concerns about environment and development. It is not only the best known and most commonly cited idea linking environment and development, it is also the best worked-out, in that it is the capstone of the World Conservation Strategy and the Brundtland Report. These documents are the subject of the next chapter, which discusses their arguments, and assesses the nature of the ideology that shapes their ideas about development. However, the concept of sustainable development cannot be understood in a historical vacuum. It has many antecedents, and over time has taken on board many accretions and influences.

The history of thinking about sustainable development is closely linked to the history of environmental concern and peoples' attitudes to nature. Both represent responses to changing scientific understanding, changing knowledge about the world and ideas about society. Where they differ, and more particularly where their histories differ, is in their geographical scope. Histories of environmentalism have tended to focus on Europe or America. However the history of sustainable development thinking must embrace the way these essentially metropolitan ideas were expressed on the periphery in the present century, initially on the colonial periphery and latterly within the countries of the independent Third World. This focuses attention in particular on the rise of international conservation and global thinking about the environment (see Boardman 1981). Globalism was a major feature of the environmentalism of the 1960s and 1970s. It was a major factor in the integration of environmentalist and developmental thinking, and hence in the forging of the specific formulations of sustainable development which are discussed in Chapter 3. Concepts of sustainable development formed an inextricably part of the environmentalism which emerged in Europe and North America in the 1960s and 1970s. The phenomenon of that emergence is well described elsewhere (e.g. O'Riordan 1976a, Sandbach 1980, Petulla 1980, Cotgrove 1982, Lowe and Goyder 1983, Pepper 1984, McCormick 1986a, 1989, Hays 1987). Its intellectual roots lie a great deal further back and are beyond the scope of this book to unravel. Fortunately they too have been explored, for example by Passmore (1980), Doughty (1981), Attfield (1983), Thomas (1983), Merchant (1980), and Pepper (1984).

The attempt to write about the history of sustainable development is made difficult by the eurocentric and americocentric nature of the literature on environmentalism, as well as its abundance. Furthermore, as McIntosh (1986) points out in the context of the historiography of ecological science, histories of ideas divorced from social and intellectual context are of little value. There is the further serious problem of looking for clear and simple 'roots' of ideas which relate to each other through time in a complex and fluid way, and which at any given time are held and articulated in diverse ways by different people. Ideas about nature, and particularly about ways to treat or manage nature, and ideas about change or 'development' in society, are among the most subtle and intractable. An account of the evolution of different strands of thought about sustainable development is therefore difficult. However, it is also necessary. If the nature of the thinking about sustainable development which emerged in the 1960s is to be understood, it is necessary to tease out some of the strands in the complex fabric of its past. This chapter does this by selecting five themes, overlapping in time. The first is that of nature preservation, and later conservation, in tropical countries. The second theme is the rise of tropical ecological science, particularly in Africa, and the associated growth of concepts of ecological managerialism. The third theme concerns the growing global reach of scientific concern, particularly in the form of the Man and the Biosphere programme. The fourth is the rise of perceptions of global environmental crisis, and the fifth is the way international environmental concern was focused by the 1972 Stockholm Conference. This Conference formed the immediate frame for the development of the ideas of sustainable development in the World Conservation Strategy.

What is offered in this chapter is no rich history of environmental thought in any particular part of the global periphery. Those histories, for example of the impact of colonial ideas about science and environment, can be written (see for example the debate on desertification in Chapter 5), but they require a far more detailed and particular approach than is possible here. Rather, this book offers a preliminary sketch of some of the more important themes in sustainable development thinking within a roughly chronological framework.

Nature preservation

In many ways wildlife or nature conservation has been the most deep-seated root of sustainable development thinking. Indeed as will be discussed below, it can be argued that notions of sustainable development were adopted partly as a means of promoting nature preservation and conservation. The history of nature conservation in countries of the industrial metropole, for example in Britain (Sheail 1976, 1981, Allen 1976, Lowe 1983) or the USA (Nash 1973, Worster 1985, Hays 1959, 1987, Frome 1984, Petulla 1980), is well established. Although intellectual roots lie much deeper and much further back (Thomas 1983, Pepper 1984), formal organization began in the nineteenth century. Thus in Britain the second half of the century saw legislation for the protection of seabirds and the establishment of a number of conservation organizations such as the Commons Open Spaces and Footpaths Preservation Society in 1865, the Royal Society for the Protection of Birds, founded initially in 1893, and the National Trust for Places of Historic Interest and Natural Beauty in 1894. In the USA the Yellowstone National Park was established in 1872, the Boone and Crocket Club formed in 1887, and the Sierra Club in 1892. A number of National Parks were established in the 1880s and 1890s in Canada, South Australia, and New Zealand (Fitter 1978). International Ornithological Congresses were held in Vienna in 1884 and Budapest in 1895, leading eventually to a treaty signed in 1902 to protect bird species 'useful in agriculture' (Boardman 1981, p. 28).

From this beginning a diverse range of initiatives grew, for example the establishment in Britain of the Society for the Promotion of Nature Reserves in 1912 and the prosecution of a protracted debate about the need for National Parks (Sheail 1975, 1976, 1981), the foundation in 1909 of the Swiss League for the Protection of Nature (primarily to raise funds for a National Park, achieved in 1914), and of the Swedish Society for the Protection of Nature. There were parallel developments in Germany (Conwentz 1914). These developments were primarily aimed at promoting the protection of nature within the industrialized nations themselves.

However, from an early date there was also concern about conservation on a wider geographical scale, in imperial or colonial possessions. Indeed, Richard Grove (1987) argues that many of the ideas current in the colonial metropoles (specifically Britain) in the later nineteenth century were both conceived and applied first in overseas territories. Concern about the depletion of forests in the Cape Colony developed in the early nineteenth century. This, with pressure for government money for the botanic garden (*inter alia* from the Royal Botanic Gardens at Kew, themselves taken over by the government in 1820, Worthington 1983) led to the appointment of a Colonial Botanist from 1858 to 1866. Legislation to preserve open areas close to Cape Town were passed in 1846, and further Acts for the preservation of forests (1959) and game (1886) followed (Grove 1987, Mackenzie 1987). Grove argues that by about 1880, a pattern of conservation derived from a mix of Indian and Cape Colony philosophies was established in southern Africa. These ideas were important not just because they were worked out there before the USA or Europe, but because their utilitarian basis brought them into conflict with settler interests. Holistic conservation ideas could not be reconciled with 'the driving interests of local European capital' (Grove 1987, p. 36). In India the state was able to stand against these forces, but not in Africa, where conservation moved away to an obsession with parks and big game, 'nimrodic obsessions', and land alienation strategies (Grove 1987, p. 37).

Certainly the main impact of environmental concern in nineteenth-century Africa grew from the hunting which accompanied the expansion of European trade and missionary work. Mackenzie (1987) identifies three phases in the extension of European hunting in Africa. The first was commercial hunting for ivory and skins, and it was through the contacts with white hunters in the 1850s that African rulers first began 'riding the tiger of European advance' (Mackenzie 1987, p. 43). This grew into hunting as a subsidy for European advance: meat for railway construction workers, or to feed and finance trade and missionary activity. In time this gave way to what Mackenzie sees as a ritualized and idealized 'hunt', with its obsession with trophies, sportsmanship and other ideals of British boys' education. As rifles improved and the number of hunters rose, astonishing numbers of game animals were killed, their deaths lovingly chronicled and recorded. As Graham picturesquely comments, 'The swirling torrents of bloodlust that were gratified are beyond our powers of measurement. Sport hunting of real wild animals will never see its like again' (Graham 1973, p. 54). By the last decades of the nineteenth century substantial areas of southern Africa were more or less emptied of game, certainly near white settlements, railways and wagon trails.

One aspect of the phenomenon of 'the hunt' was the growth in interest in natural history and 'pseudo-science' (Mackenzie 1987) such as specimen and trophy hunting and taxonomy. This led to the emergence of ideas of controlling hunting, and hence eventually ideas of conservation. Such a move parallels the evolution in Britain from the collection of natural history specimens to conservation (Allen 1976). Theodore Roosevelt perhaps personifies this best. An avid hunter and collector of trophies, he was by the early years of this century entering enthusiastically into the debate in the USA about conservation, taking the utilitarian line of the forester Pinchot against the more romantic notions of John Muir. He subsequently became, in the words of R.S.R. Fitter, one of the 'pioneer statesmen of the movement' (Fitter 1978, p. 16).

The most obvious aspect of the conservation based on this hunting ethos was the complete denial of hunting to Africans. White men hunted; Africans poached. This denial was achieved through control on firearms, and latterly by the establishment of game reserves. The Cape Act for the Preservation of Game was passed in 1886 and extended to the British South African Territories in 1891 (Mackenzie 1987). In 1892 the Sabie Game Reserve was established in the Transvaal (to become the Kruger National Park in 1926), and in 1899 a game reserve was established in Kenya enclosing the present Amboseli National Park (Lindsay 1987). In 1900 the Kenyan Game Ordinance was passed effectively banning all hunting except by licence (Graham 1973).

The work of Game departments in Africa varied both between territories and over time, as Graham's critique of the Kenyan and Ugandan Game Departments shows (Graham 1973). Nonetheless the 'poaching problem' was a regularly repeated motif. The classic view on the destructiveness of African hunting ran as follows:

> It is commonly thought that the visiting sportsman is responsible for the decline of the African fauna. That is not so. The sportsman does not obliterate wild life. True, he kills. But seldom is the killing wholesale or indiscriminate. What the sportsman wants is a good trophy, almost invariably a male trophy, and the getting of that usually satisfies him. ... The position is not the same with the native hunter. He cares nothing about species or trophies or sex, nor does he hunt for the fun of the thing. What the native wants is as many animals as possible for the purpose either of meat or barter.
>
> (Hingston 1931, p. 404)

This particular formulation appeared at a later period, between the two wars, but such views were typical of hunter-conservationists at the turn of the century. They remained an important element in thinking about wildlife in developing countries. The importance of such ideas in the establishment of reserves for wildlife conservation is considerable (see Chapter 8). The paternalism and ideological blindness of these views (or perhaps more accurately the pretence at a 'scientific' view which is beyond ideology) as well as the sheer political unacceptability of imposed rules for environmental management resurface in debates about environmentalism and development.

Hunting in Africa also led to significant institutional developments. A conference of African colonial powers (Britain, Germany, France, Portugal, Spain, Italy and the Belgian Congo) met in London in 1900 and signed a Convention for the Preservation of Animals, Birds and Fish in Africa (Fitter 1978). Three years later the threat of de-gazetting of a game reserve north of the River Sobat in the Sudan led to the establishment of the Society for the Preservation of the Wild Fauna of the Empire, with a powerful list of ordinary and honorary members (including Kitchener of Khartoum, President Theodore Roosevelt and the British Secretary of State of the Colonies; Fitter 1978). The 'penitent butchers', as they became nicknamed, helped place preservation on the agenda of colonial management. The 1900 Convention was never set in operation, but in the early decades of the century there were developments in a number of other colonial nations. For example, in the mid-1920s a French Permanent Committee for the Protection of Colonial Fauna was established, and in 1925 King Albert created the Parc National Albert (now the Virunga National Park), the first African National Park (Fitter 1978, Boardman 1981).

More interesting in many ways was the pressure growing in industrialized countries, primarily in Western Europe for an international organization to promote conservation (Boardman 1981). The idea was mooted at an International Congress for the Preservation of Nature held in Paris in 1909, and in fact an Act of Foundation of a Consultative Commission for the International Protection of Nature was signed at Berne in 1913 by delegates from seventeen European countries. However, war prevented any substantive action, and despite the recommendation of another Congress at Paris in 1923, it was not until the assembly of the International Union of the Biological Sciences in 1928 that an Office International de Documentation et de Corrélation pour la Protection de la Nature was established (Boardman 1981). This was consolidated into the International Office for the Protection of Nature (IOPN) in 1934. The IOPN joined the International Committee for Bird Protection (ICBP, now the International Council for Bird Preservation) formed in 1922 to promote nature protection on a global scale.

Concern for nature in colonial territories, particularly Africa, persisted, and in the 1930s the IOPN published a series of reports on African colonial territories. In this they were particularly supported by

American interest. Following President Roosevelt's lead, American concern about Africa, 'that region where scientists and naturalists felt the greatest threats to the world's fauna and flora existed' (Boardman 1981), grew, although it remained fragmentary until after the Second World War. In 1930 the Boone and Crockett Club set up the American Committee for International Wildlife Protection. The ACIWLP assisted the IOPN financially and also itself promoted nature protection and carried out research, particularly during the war when the work of IOPN (based in Amsterdam) was severely disrupted.

Africa also remained the target of pre-war concern by the Society for the Protection of the Fauna of the Empire (SPFE: the word 'wild' was dropped from their title after the First World War). For example, Hingston's visit to Africa was at the request of the SPFE, and his paper to the Royal Geographical Society on the need for National Parks was plain and persuasive: 'It is as certain as night follows day that unless vigorous and adequate precautions be taken several of the largest mammals of Africa will within the next two or three decades become totally extinct' (Hingston 1931, p. 402). Following the International Congress for the Protection of Nature in Paris in 1931, a new intergovernmental conference was called for 1933 (Boardman 1981).

After the war, efforts to strengthen international nature protection were renewed through both the IOPN and the ICBP, and the Swiss League for the Protection of Nature. The idea of a new organization was mooted at a conference in Basle in 1946, and taken up by the newly established UNESCO (the United Nations Educational, Scientific and Cultural Organization) in the person of its first Director, Julian Huxley (Huxley 1977). Huxley had been chairman of the Wild Life Conservation Special Committee (England and Wales) which in 1947 had recommended government involvement in nature conservation in Britain and the establishment of a Biological Service (later substantially implemented in the ecreation of the Nature Conservancy) (Sheail 1975, 1984a). Huxley also had long-standing interests in Africa, for example as a member of the African Survey Research Committee formed in Britain in the early 1930s (Huxley 1930, Worthington 1983). The UNESCO General Conferences in 1946 and 1947 were persuaded to take nature conservation on board, and a Provisional Union for the Protection of Nature was set up at a further meeting at Brunnen the same year. Political difficulties abounded, but a constitution was adopted at a joint French Government/UNESCO Conference at Fontainebleau in 1948. UNESCO granted financial support a month later. The following year the new International Union for the Protection of Nature (IUPN) launched a 'trailblazing endeavour' (Boardman 1981, p. 51) at the Lake Success Conference in August 1949, which UNESCO had arranged to follow the United Nations

Conference on the Conservation and Utilization of Resources in New York.

Despite constraints of funding, the IUPN survived and prospered, changing its name in 1956 to the International Union for Conservation of Nature (IUCN). From the first, its focus was firmly placed wider than the industrialized countries of the West. Data collection on endangered species began, and was strengthened by extra funding after 1955 which put a biologist into the field in Africa, South Asia and the Middle East to report on threatened mammals. The Red Data Books, as they appeared from 1966, contained many species from developing countries. Interest in the tropics, and still particularly in Africa, was maintained also by other organizations, for example the Frankfurt Zoological Society through the work of Bernard Grzimek in Kenya and Tanzania, and the Conservation Foundation (set up in 1948 by the New York Zoological Society under Fairfield Osborn) through Fraser Darling.

IUCN itself launched several projects in the early 1960s which had specific reference to the emerging 'developing world'. The MAR project on wetlands was launched, the Ramsar Convention eventually being signed in 1971 in Iran. More relevant, perhaps, was the African Special Project (1961-3). Concern about African wildlife was a feature of both the Fontainebleau meeting in 1948 and the Lake Success Conference the following year. Several National Parks were declared in colonial territories in the late 1940s, for example the Nairobi National Park in Kenya in 1946 and Tsavo two years later, Serengeti in Tanganyika in 1951 and Murchison Falls in Uganda in 1951 (Fitter 1978). In 1953 a special conference on African conservation problems convened at Bukavu in the Belgian Congo. These developments did not end concern, and by the IUCN's General Assembly in 1960 Africa 'had emerged as the central problem overshadowing all else' (Boardman 1981, p. 148). Africa was becoming independent, and political control was shifting away from the metropole; the poachers were turning gamekeepers, and might not follow the same policies. IUCN therefore joined with the FAO to launch the African Special Project to influence public opinion in Africa itself.

The centrepiece of the African Special Project was the conference at Arusha in Tanzania in 1961. Much was made of the 'Arusha Declaration on Conservation', a telegram sent by Julius Nyerere (and not to be confused with the Arusha Declaration of 1967 when Nyerere launched Tanzania into 'African Socialism'). This represented a strong personal statement of commitment to wildlife conservation, but also stressed the Project's focus on wider concerns of resource development:

> The survival of our wildlife is a matter of grave concern to all of us in Africa. These wild creatures amid the wild places they inhabit are not only important as a source of wonder and inspiration but are an integral part of our natural resources and of our future livelihood and well-being.
>
> (Worthington 1983, p. 154).

The African Special Project was followed up by IUCN missions to seventeen African countries (Fitter 1978), and eventually (after considerable political haggling between the IUCN and FAO) a new African Convention on the Conservation of Nature and Natural Resources was adopted by the Organization of African Unity in Addis Ababa in 1968 (Boardman 1981).

Not all attention focused on Africa. IUCN was involved in the establishment of the Charles Darwin Foundation for the Galapagos Islands in Ecuador in 1959, and the Fauna Preservation Society in 'Operation Oryx' in the Middle East in 1961-2. Further conferences were also held in Bangkok in 1965 and at San Carlos de Bariloche in Argentina in 1968 to try to recreate the effect of Arusha (Fitter 1978). IUCN also made a series of approaches to governments in the early years about the status of particular rare species, for example Indonesia (orang-utang and elephant), and Nepal and India (rhino). However, these had limited success, often being greeted by 'polite rebuffs, disguised either as bland encouragement or the forwarding of notes to other departments' (Boardman 1981, p. 76).

One reason for this was IUCN's eurocentric and 'First World' image, which continued to limit its credibility in the Third World. Unfortunately, as Boardman comments, 'Western Europe and North America were the habitat of the conservationist' (p. 114). Thus although by 1976 twenty African countries were represented by government agencies or non-governmental organizations as members of IUCN, they were swamped by the number of First World members: 57 of the 244 NGO members came from the USA. This relative domination by First World organizations has led to periodic complaints (for example by the Kenyan Delegation at the General Assembly in New Delhi in 1969; Boardman 1981), and indeed, Boardman suggests that it was one factor in the development by IUCN of ideas, such as the African Special Project, that embraced development. Such initiatives were important forerunners of the idea of 'sustainable development' contained in the World Conservation Strategy. The emergence of this document is discussed later in this book. However, first it is necessary to backtrack slightly to examine other inputs into concepts of sustainable development, starting with the role of science and scientific thinking in views of tropical environments.

Tropical ecology and managerialism

The science of ecology developed at the end of the nineteenth century in Europe and the USA (Lowe 1976, Kormondy and McCormick 1979, Worster 1985, McIntosh 1985, Sheail 1987). There were close links from an early date between the new science and the preservation or conservation movement, particularly in the UK (Sheail 1976, 1987, Lowe 1983). Ecology was also closely associated with the rise of environmentalism in the 1970s. At this period (and to a lesser extent today) the word ecology was used widely even where ideas owed little or nothing to scientific ideas or method (Enzensberger 1974, Lowe and Warboys 1975, 1976, Ridgeway 1971), but even then it was true that many of the prominent figures of the environmental movement were trained in ecology (Chisholm 1972).

Among many other attributes, ecology has at different times seemed to offer new, value-free and apolitical ways of not only understanding but also managing the environment. This ecological 'managerialism' was particularly attractive in places such as Africa at the end of the Second World War. The natural environment seemed relatively little affected by people, and there were few effective models for the burgeoning task of development planning. Thus the acquisition of scientific understanding of tropical environments was followed in time by a concern for the application of that knowledge to development. This utility of ecological science paralleled the rising demand for conservation and formed a second strand in the growth of sustainable development thinking.

The tropics have been an important hearth of innovation in biological science, and particularly ecology. Indeed, Goodland (1975) argues that the tropics have played a pivotal role in the development of ecology, choosing as a hook on which to hang his argument the occasion of Eugen Warming's Jubilee. Certainly the attention of the practitioners of 'self-conscious ecology' (McIntosh 1986) in Europe and North America was soon extended to the study of the diversity of nature overseas. Indeed this was an extension of the close and reciprocal links which had existed between science and exploration through the eighteenth and nineteenth centuries, with environmental research both a feature of and stimulus to exploration of the Tropics (Stoddart 1986, see also Crosby 1986). It is possible to outline something of the scope of the development of ecology and ecological managerialism in the tropics using examples from Britain and anglophone Africa. However, obviously this is far from a full account of developments in the tropics as a whole, and the story elsewhere may vary considerably.

When the British Ecological Society was founded in 1914, growing out of the British Vegetation Committee (Salisbury 1964, Pearsall

1964, Allen 1976, 1986), the first volume of the new Journal of Ecology in that year contained reviews of publications on the forests of British Guiana (Anderson 1912) and studies of vegetation in Natal and the eastern Himalayas. The second volume reviewed works on the vegetation of Aden, the highlands of North Borneo and Sikkim, and the following year added work on the dipterocarp forests of the Philippines and montane rain forest in Jamaica.

Interest in the ecology, and particularly the vegetation, outside Britain was maintained, and in Volume 12 (1924) a specific section was instituted quaintly entitled 'notices of publications on foreign vegetation'. The majority of publications reviewed concerned the USA, Australia and South Africa, but there were others, for example Cuba (Volume 15, 1927). The journal also in time carried substantive papers on the vegetation of the Empire. Perhaps unsurprisingly, the first full paper concerned the vegetation of South Africa (Bews 1916), but this was followed by a steady trickle of papers on various parts of the world, both inside and outside the Empire, for example work on the rain forest of South Brazil (McLean 1919) and the forests of the Garhwal Himalaya (Osmaston 1922). Nor was such work confined to British ecologists. Studies on the vegetation and soils of Africa were begun in 1918 at the instigation of the American Commission to Negotiate Peace, involving a journey from Cape to Cairo from 1919 to 1920 and a monograph on African vegetation, soils and land classification (Shantz and Marbut 1923).

The later 1920s saw British ecologists working on the ground in a number of tropical locations. For example, there were expeditions from Oxford to British Guiana in 1929 (Hingston 1930, Davis and Richards 1933) and to Sarawak in 1932 (Harrison 1933), two expeditions to the East African Lakes (Worthington 1932, 1983) and a Cambridge Botanical Expedition to Nigeria in 1935 (Richards 1939, Evans 1939). By that time ecological research on disease in Africa was well established, notably on the tsetse fly (Buxton 1935, Huxley 1930, Ford 1971). Biological science was readily harnessed in the service of colonial development. In 1928, for example, the Journal of Ecology carried a paper by the Assistant Economic Botanist in the Bombay Department of Agriculture on the aquatic weeds of irrigation canals in the Deccan (Narayanayaya 1928), and the geographer Dudley Stamp described an aerial survey of the Irrawaddy Delta forests (Stamp 1925). In reviewing work on the East African Lakes many years later, E.B. Worthington comments: 'Advice on the existing and potential fisheries was a primary objective, but to provide this it was necessary to elucidate the ecology of each lake' (Worthington 1983, p. 13). This was no isolated instance of ecologists' interests in essentially economic problems.

The Imperial Botanical Conference, held at Imperial College in London in 1924, stressed the need for 'a complete botanical survey of the different parts of the Empire' (Brooks 1925, p. 156). This conference was the successor to the International Botanical Congresses in Paris (1900), Vienna (1905) and Brussels (1910), and was deferred until the British Empire Exhibition took place. Science was to serve Imperial commerce: ' In such a Survey the Imperial Government has good reason to take a prominent part, as it depends so much on the overseas portions of the Empire for the supply of raw materials for manufacture, and of foodstuffs' (Davy 1925, p. 215). Throughout, the tone of the Conference was strongly economic and utilitarian: 'I submit that it is our duty as botanists to enlighten the world of commerce, as far as may lie in our power, with regard to plants in their relation to man and their relation to conditions of soil and climate' (Hill 1925, p. 198). The conference passed a resolution establishing the British Empire Vegetation Committee. Under the Chairmanship of A.G. Tansley (first Editor of the Journal of Ecology), this published a monograph, *Aims and Methods in the Study of Vegetation*, in 1926. However, a further twelve years passed before the first in a planned series of regional monographs appeared, *The Vegetation of South Africa* (Adamson 1938). Ecology, however, was seen to have more to offer than simply the description of vegetation. Shantz and Marbut suggested that when vegetation was 'properly analyzed and classified' it could serve 'a most practical purpose'. Vegetation survey could allow the definition of natural regions 'for use in the development of both agriculture and forestry' (Shantz and Marbut 1923, p. 4). Phillips (1931) advocated a progressive scheme of ecological investigation in East Central and South Africa, including detailed suggestions on staffing, research programme and administrative arrangements. This was a truly catholic vision of the contribution of ecology to the use of resources. His objects included 'the rational use of biotic communities in agricultural practice', the 'wise utilization of natural grazing', the development of 'progressive forestry policies' through 'conservation, re-forestation, afforestation, sylvicultural management and forest production', the 'prevention of soil erosion and its concomitant evils', and catchment water conservation and research into tsetse fly (Phillips 1931, p. 474).

This approach was to be expanded in the work of the African Survey (Hailey 1938), carried out under the aegis of an African Research Survey Committee, on which the two biologists were Julian Huxley (later Director of UNESCO) and John Orr (later Director of FAO). In 1935, Sir Malcolm Hailey (later Lord Hailey) drove in two Fords from Cape Town to the Mediterranean. With him for part of the way went E.B. Worthington, who wrote a contributory volume *Science in Africa*

(Worthington 1938). This report was influential, both on the main volume and elsewhere. It took the utilitarian line that science (particularly ecology) was useful in development (or 'bonification' as Worthington called 'the promotion of human welfare'):

> a key problem was how *Homo sapiens* could himself benefit from this vast ecological complex which was Africa, how he could live and multiply on the income of the natural resources without destroying their capital ... and how he could conserve the values of Africa for future generations, not only the economic values but also the scientific and ethical values.
>
> (Worthington 1983, p. 46)

Science in Africa broke new ground. Not only did it place science in the service of a loosely-defined 'development', but it also used an explicitly ecological structure in presentation. The report moved from geology and meteorology through botany and zoology to agriculture, fisheries, disease, population and anthropology, and stressed the fact that interrelations between the sciences had 'important practical applications' (Worthington 1938, p. 3). This approach was developed in the post-war period. The Conseil Scientifique pour L'Afrique (CSA), established following an Empire Scientific Conference held by the Royal Society in London in 1946 and a further conference in Johannesburg in 1949, developed similar arguments about the importance of science to development (Worthington 1958). Increased interest in scientific research in African territories at this time saw the establishment of the British Colonial Research Council (under Lord Hailey), the French Office de Récherche Scientifique et Technique d'Outre Mer (ORSTOM), and the Belgian Institut pour la Récherche Scientifique en Afrique Central (IRSAC) which joined the existing agricultural body, Institut National pour l'Étude Agronomique du Congo Belge (INEAC) (Worthington 1983).

In post-war Africa, as development planning began to grow, ecology was seen to provide not only valuable data, but a model. Thus in a remarkable polemic in support of a managerial role for environmental science, Culwick (1943) suggested that development in Tanganyika was essentially a problem of 'controlling the environment and exploiting it in such a way that not only this but succeeding generations may be able to attain their full potentialities, both physical and mental'. This task of development was 'primarily a scientific problem in which all branches of science, physical and social, have an essential part to play' (Culwick 1943, p. 1). Furthermore, the task would require reform of government, so that 'our post-war life can be planned as an oecological whole'. In the past, scientists had been regarded as 'appendages to the main administrative body, called in every now and again for consultation when a scientific question cropped up', but now the running of the Territory had to be regarded 'primarily as a scientific affair', and administration should therefore become 'merely the mechanism for putting the big scientific plan into action' (Culwick 1943, p. 5).

The administration in Tanganyika never reduced itself to that, but certainly what Low and Lonsdale (1976) have called a 'second colonial occupation' in East Africa involved a new and important role for scientists in some fields, most notably perhaps agriculturalists. Furthermore ecological managerialism did occasionally penetrate the development planning process. Ecological ideas were given an important position in the First Ugandan Development Plan, which the new Governor in 1946 thought was 'much too departmental' (Worthington 1983, p. 92). In this it was not unusual, the scientific journal *Nature* complained in 1948 of one plan which was 'an agglomeration of departmental suggestions masquerading as a development plan' (*Nature* 1948). The model being offered in this territory, which depended to such a large extent on living resources, was explicitly ecological: 'Fundamentally, the problem of development was one of human ecology, the diverse people reacting with their varied environments' (Worthington 1983, p. 97). This approach to economic development of human ecology was not unique (a famous example being Frank Fraser Darling's West Highland Survey; Fraser Darling 1955), but it represents a significant element in the development of environmentalist thinking about the development process.

Environmentalism and global crisis

The rise of environmentalism in the industrialized world in the 1960s had enormous significance for debates about the role of ecology and conservation in development. Although Myers and Myers (1982) argue that, even in the 1980s, perception of environmental issues rarely transcends national boundaries or national interests, one of the most distinctive features of the environmentalism of the 1960s and 1970s was its concern for the global environment issues. By the 1970s the notion of the 'Spaceship Earth', the term used by Boulding (1966), had become commonplace. It is perfectly summed up in the title of the book written in preparation for the 1972 UN Conference on the Human Environment in Stockholm, *Only One Earth* (Ward and Dubos 1972). Partly through the Stockholm Conference, crisis thinking on a global scale became a key theme in the environmental revolution or 'new environmentalism' (Cotgrove 1982) of Europe and North America in the 1970s. Parallel with it grew an apocalyptic vision of neo-Malthusian crisis. It is within this context that ideas about sustainable development emerged.

Accounts of the nature and growth of environmentalism are usually confined to the industrialized world, particularly Europe and North America (although see Capra and Spretnak 1984). Indeed, Redclift argues that its ethnocentric nature should 'make us wary of international comparisons based entirely on European or North American experience' (Redclift 1984, p. 46). In fact, from the first, this First World environmentalism was based on a fundamentally global view. The 'doomsday syndrome' attacked by Maddox (1972) was essentially global in scope. This globalism embraced both the recognition of environmental problems that transcended national boundaries, notably pollution (Dahlberg *et al.* 1985), and the spread of concern about the environment outside industrialized countries.

Such globalism in environmental concern is not new. In 1954, Fairfield Osborn wrote: 'man is becoming aware of the limits of his earth. The isolation of a nation, or even a tribe, is a condition of an age gone by (Osborn 1954, p. 11). One fruit of this realization was the rise of concern about global population. Ecologists such as Raymond Pearl discussed the phenomenon of human population growth in the 1920s (Pearl 1927), and the spectre of population growth was raised by commentators such as Carr-Saunders (1922, 1936). In 1948, Osborn commented: 'the tide of the earth's population is rising, the reservoir of the earth's living resources is falling' (Osborn 1948, p. 68).

Under Julian Huxley, UNESCO became involved in the 1950s in the application of scientific evidence in the debate about population and development. For example UNESCO ran the World Population Conference with FAO in Rome in 1954. As such it linked development issues to one of the central concerns of environmentalism. The 'population problem', as it became widely known, was commented on extensively in the 1950s, for example by Boyd-Orr (1953), Stamp (1953), Russell (1954) and in the famous conference *Man's Role in Changing the Face of the Earth* (Thomas 1956). Neo-Malthusian arguments which were a prominent feature of environmentalism in the 1960s and 1970s (e.g. Ehrlich and Ehrlich 1970, and the starkly apocalyptic *The Population Bomb*, Ehrlich 1972) had already been rehearsed in a colonial (and developmental) context some decades before. Thus *Nature* commented in 1952, 'If the human race continues to multiply at the present rate, catastrophe will only be avoided if methods of cultivation are used which, far from destroying the fertility of the soil, maintain or even increase it (Nature 1952, pp. 677-8).

The scope of neo-Malthusian thinking has changed very little since the 1970s. In 1970 the world population was 3.5 billion, and

'Spaceship Earth' was said to be 'filled to capacity and beyond and is running out of food' (Ehrlich and Ehrlich 1970, p. 3). In July 1987 global population had risen, according to estimates by the United Nations Fund of Population Affairs, to 5 billion people, and yet in the 1980s neither the analysis nor the rhetoric of neo-Malthusianism had changed: 'whichever way one looks at the population problem - whether as a biologist, sociologist, theologian, medical doctor, industrialist, administrator or politician - it is obvious that it presents the greatest menace to the future of the biosphere' (Worthington 1982a, p. 98). To some environmentalists, population growth continues to offer a dire and global threat: 'the remedy is left to Nature's ways of shortage and deprivation, famine or pestilence, and to Mankind's own way of increasing violence and slaughter' (Polunin 1984, p. 296) The 'catastrophist' environmentalist thinking of the 1960s and 1970s also embraced a parallel and closely related debate about economic growth. Over the 1960s and 1970s economic growth in industrialized countries gave way to recession and inflation. Most analysts followed Kahn and Wiener (1967) who offered a fairly optimistic range of scenarios of the future. Within a few years the tone of forecasts had changed. In 1969 Mishan looked at the costs of economic growth (coining the delightful term 'growthmania'), and concluded rather cautiously that 'the continued pursuit of economic growth by Western societies is more likely on balance to reduce rather than increase social welfare'.

This still cautious view was overtaken by the results of attempts to produce global computer models of the 'world system', notably Forrester's *World Dynamics* (1971). This approach was developed by a team from the Massachussetts Institute of Technology for the 'Club of Rome', an international group set up with the backing of European multinational companies (Golub and Townsend 1977). The work was published as *The Limits to Growth* (Meadows *et al.* 1972), and with the British utopian polemic *Blueprint for Survival* published in the same year (Goldsmith *et al.*1972), it forms one of the two most commonly quoted (although perhaps less commonly read) treatises of 1970s environmentalism.

These volumes are widely quoted and reviewed (e.g. O'Riordan 1981, Pepper 1984) and need not be described in detail here. The basic idea of zero growth or the steady state economy have attracted sober support (e.g. Daly 1973, 1977, Mishan 1977), but they have also generated a body of fairly vituperative criticism. *The Limits to Growth* has been the more marked target. Beckerman (1974) suggests the Club of Rome report is 'guilty of various kinds of flagrant errors of fact, logic and scientific method' (p. 242), while Simon (1981) dismisses it as 'a fascinating example of how scientific work can be outrageously bad and yet be very influential' (p. 286). Maddox (1972) says simply 'the doomsday cause would be more telling if it were more securely grounded in facts' (p. 2).

Neo-Malthusian thinking, both about growth and more particularly about population, has significant political implications. Some environmentalists seem to despair of existing political structures for change, and call for a technocratic global government. Thus Myers and Myers (1982) discuss the notion of technocratic political globalism, with supranational power exercised by some notional 'global community', and Polunin suggests the world 'needs saving from itself - from destruction perpetrated by Mankind, its uniquely intelligent component' (Polunin 1984, pp. 294-5). One example of such environmentalist thinking about 'apolitical' globalism is the 'World Campaign for the Biosphere' launched in 1982 (Worthington 1982a, *Environmental Conservation* 1982). The Campaign's Council, composed of 'a group of internationally recognised environmental leaders in science, policy and business' is dedicated to 'understanding and protecting the Biosphere' (Baldwin and Sacks 1984).

This 'apoliticism' is obviously not apolitical at all, and indeed there are powerful radical critiques of neo-Malthusianism and systems theory (e.g. Enzensberger 1974, Harvey 1974, Gregory 1980). These are discussed in more detail in Chapter 4. The important point to note at this stage is the way the notion of 'global crisis' in the 1960s and 1970s contributed to the internationalization of both ecology and environmentalism. Although the 'ecological prescriptions for managing the human use of the earth' which were on offer were indeed extremely limited (Stoddart 1970, p. 2), the grandiose claims and global fears of environmentalism were far from ineffective.

Global ecology, conservation and the environment

Theoretical and practical links between ecological science and development were fostered in 1964 by the establishment of the International Biological Programme (IBP). The IBP was launched by the International Union of Biological Sciences (IUBS) under the International Council of Scientific Unions (ICSU) in 1964, the model of international scientific cooperation initially coming from the International Geophysical Year 1957-8 (Worthington 1975, 1982a, b, 1983). Planning began in 1959, with the vision of studying 'the biological basis of man's welfare' (Waddington 1975, p. 5). However, it was only at a meeting at the IUCN headquarters in Morges in 1962 that the IBP planning committee adopted the sevenfold grouping of research which came to form the structure of the programme. Perhaps unsurprisingly given the venue, conservation formed one group, the 'Conservation Terrestrial' (CT) section, (Nicholson 1975).

In the event, extensive work was done by the CT Section, in particular the gathering of global data on areas of scientific importance, and an attempt to establish a network of biome research stations. However, Nicholson argues that the biological community 'never fully endorsed in practise the inclusion of conservation' (Nicholson 1975, p. 14). There were also substantial problems of integration. Boffey (1976), for example, points out that the level of cooperation between First World scientists and their Third World counterparts was small. The CT Section faced particular problems. It had to be Janus-headed, and address two audiences, both research biologists, to encourage them to apply their ideas, and natural resource managers, to encourage the application of ecological theory (Worthington 1975, p. 86). The results were mixed.

The IBP, however, was only one element in the expansion of research into the global environment, and awareness of global environmental problems. There were a series of meetings called in the post-war period to 'bring together key actors of the globe' to address environmental issues on a global scale (Dahlberg *et al.* 1985). In some ways the most fundamental was a symposium held by the Wenner-Gren Foundation at Princeton New Jersey in 1955, *Man's Role in Changing the Face of the Earth* (Thomas 1956), which was chaired by Carl Sauer, Marston Bates and Lewis Mumford. This symposium claimed roots in the work of George Perkins Marsh (1864) on the influence of man on nature, and similar thinkers from the late nineteenth century onwards. Man, 'the ecological dominant on this planet', needed the insights of scholars to understand his 'impress' on the earth. The symposium was 'a first attempt to provide an integrated basis for such an insight and to demonstrate the capacity of a great number of fields of knowledge to add to our understanding' (Thomas 1956, p. xxxvii). This theme was one which fitted easily into the new internationalism of science in the IBP, and with the rising globalism of environmentalist thinking in the 1960s.

The global nature of environmental problems was debated at the Third General Assembly of the IBP at Varna in Bulgaria in April 1968, and in 1969 the Scientific Committee for Problems of the Environment (SCOPE) was established under the International Council of Scientific Unions (ICSU) (Worthington 1982b, 1983). SCOPE's work has been, to use O'Riordan's classification, strongly technocentric, focusing on ways of studying, understanding specific environmental problems, particularly at a global scale. Thus there have been reports in 1971 and 1973 on global environmental monitoring, in

1975 and 1979 on global geochemical cycles, and in 1979 on Saharan dust. SCOPE's work also collated material on specific environmental issues, for example environmental impact assessment (Munn 1979).

To an extent the work of SCOPE was overtaken (although not replaced) by the Man and the Biosphere Programme which grew out of the 'Biosphere Conference', the *Intergovernmental Conference of Experts on a Scientific Basis for Rational Use and Conservation of the Biosphere* held in Paris in 1968. This made explicit the growing engagement by conservationists and environmentalists with the development process (Caldwell 1984). It was a further step in the incorporation of Third World countries into the world of international environmental concern. The initiative stemmed primarily from a realization of the continuing failure to create a truly international environmentalism. Boardman comments that 'in the last analysis, the one major political cleavage that the issue of nature conservation has failed to bridge adequately is that between industrialized and developing countries' (Boardman 1981, p. 19). Conservationists within IUCN and organizations such as UNESCO realized increasingly through the 1960s that they could not influence decisions about the use of natural resources in the Third World unless they were prepared at least to talk in the new language of development. Further to this, there was increasing recognition of the need to make conservation more 'relevant' to the needs of the emerging Third World.

IUCN progressively repackaged its message to embrace development through the 1950s and 1960s. In 1956 their name was changed to the International Union for Conservation of Nature and Natural Resources, a change which 'symbolized the conviction reached over the previous eight years that "nature", the fauna and flora of the living world, is essentially part of the living resources of the planet; it also implied that social and economic considerations must enter into the problem of conservation' (Munro 1978, p. 14). IUCN cooperated in the publication of *Ecological Principles for Economic Development* (Dasmann *et al.* 1973) following the conference at Washington University in Virginia in 1968 on ecology and international development (Farvar and Milton 1973).

UNESCO had adopted key resolutions which explicitly linked conservation and development at their 1962 General Conference. It sponsored the symposium on *Man's Place in the Island Ecosystem* at the 10th Pacific Science Congress in Honolulu in 1961 as part of the Humid Tropics Programme of its Natural Sciences department, which developed the idea of man's actions as a functioning part of an ecosystem (Fosberg 1963). UNESCO also joined other UN Agencies (FAO and UNDP), IUCN, the Conservation Foundation and the World Bank in the discussions about ecology and development which followed this conference (Dasmann *et al.* 1973), and had been involved in a review of natural resources in Africa at the request of the Economic Commission for Africa in 1959 (UNESCO 1963).

With this background it is not a surprise that the Biosphere Conference had a complicated history, starting with discussion at the IUCN General Assembly in Nairobi in 1968. IUCN ideas of an international conference on endangered species awoke little interest in ECOSOC (the UN Economic and Social Council), and UNESCO ideas of a broader conference embracing development led to the 'biosphere' approach (Boardman 1981). The conference heard accounts of human impacts on the biosphere and resource depletion, and called for the establishment of an interdisciplinary and international programme of research on the rational use of natural resources and 'to deal with global environmental problems' (Gilbert and Christy 1981). This proposal was clearly heavily influenced by the experiences with the International Biological Programme, and indeed Worthington points out that the many IBP workers in national delegations 'made much of the running' at Paris (Worthington 1983, p. 175). A draft of a 'Man and the Biosphere' (MAB) programme was approved by the UNESCO General Assembly in 1970, and MAB was launched in 1971 under an International Coordinating Council (Gilbert and Christy 1981). It was 'in large measure the successor of IBP' (Worthington 1983, p. 176). There was considerable passage of scientific information between the two programmes (Worthington 1975).

MAB's function was 'to develop the basis within the natural and social sciences for the rational use and conservation of the resources of the biosphere and for the improvement of the global relationships between man and the environment' (Gilbert and Christy 1981). As such, MAB was to be both useful and down-to-earth: 'ivory tower research' was of little use to 'those who have to make management decisions in a world of increasing complexity'. MAB would be different, breaking down 'obsolete barriers' between natural and social scientists and decision-makers, and offering instead 'an interdisciplinary, problem-oriented approach to the management of natural and man-modified ecosystems' (UNESCO not dated). MAB was given an exhaustive and astonishingly open-ended range of specific objectives. These included the study of the 'structure, functioning and dynamics of natural, modified and managed ecosystems' and of the relations between 'natural' ecosystems and 'socio-economic processes' and the identification and assessment of human impacts on the biosphere. There were also aims to promote 'global coherence of environmental research', environmental education and and specialist training, and 'global awareness of environmental problems' (Gilbert and Christy 1981, pp. 704-5).

These aims were obviously not all attainable in full, so a series of specific fields were set out for action. These focused either on particular environments of concern to environmentalists (for example rainforests, Project 1, or semi-arid zones, Project 3), or particular impacts of development (for example major engineering works, urban systems and energy, or pollution, Projects 10, 11 and 14, the latter added in 1974). The main shortcomings of the programme were due quite simply to its ambition. Government response in the Third World was favourable, but overall international action was slow to get off the ground and under-funded (Batisse 1975). There certainly were projects which became operational in the Third World. For example under MAB Project 1 on the ecological effects of increasing human activities on tropical and subtropical forest ecosystems, the San Carlos de Rio Negro project was begun in 1976 through a newly established International Center for Tropical Ecology in Caracas. Studies of the structure, composition and production of undisturbed rainforest soils were followed by research on nutrient cycling and the effects of disturbance (Gilbert and Christy 1981). Under MAB Project 3 on the impact of human activities and land use practices on grazing lands, the Integrated Project on Arid Lands (IPAL) was begun in 1976 with UNESCO and UNEP funding, and has produced a stream of integrated research studies on semi-arid vegetation and the ecology of pastoralism (e.g. Herlocker 1979). Under MAB Project 7 on island ecosystems, a pilot project was organized in Fiji from 1974 to 1976 supported by the UN Fund for Population Activities (Bayliss-Smith *et al.* 1988), and a second in selected islands of the eastern Caribbean in 1979 (di Castri 1986).

These projects, and others, have a good claim to be the forerunners of 'sustainable development' thinking, linking natural ecosystems and human use in an innovative or wholly research-based structure. However, not all MAB activities were so new in their outlook. The established concern of nature conservation readily showed through the new clothes of human ecology. The best example of this is Project 8, 'the conservation of natural areas and the genetic information they contain', through the establishment of 'biosphere reserves'. Unlike a number of the other Projects, Project 8 allowed existing nature or wildlife conservation activities to be reclassified (and sometimes extended) to fit the MAB framework. As such it also represented an important foothold for 'pure' wildlife conservation within the much broader aims of MAB, and provided a link back to nature preservation from the new fields of conservation for human use. Thus Batisse wrote 'The greatest merit of the 'Biosphere Conference' was perhaps the assertion, for the first time in an intergovernmental context, that the conservation of environmental resources could and should be achieved alongside that of their utilization for human benefit' (Batisse 1982, p. 101).

The notion of biosphere reserves was discussed early in MAB's development in 1969, and was progressively defined at the 1970 UNESCO General Conference and the 1971 MAB International Coordinating Council. The importance of linking the reserves and research, and of defining 'a coordinated worldwide network of areas' were established (Batisse 1982, p. 101). Criteria for selection and objectives were set out by the Expert Panel in 1973 (UNESCO 1973). Among these were the need to include representative examples of ecological areas and unique communities, the need for reserves to be large enough to be 'an effective conservation unit' which can accommodate different uses without conflict', and the importance of providing sites for ecological research (Batisse 1982, p. 102). These criteria resemble closely those developed in Britain for the selection of National Nature Reserves and Sites of Special Scientific Interest: representativeness, rarity, extent and recorded history (Ratcliffe 1971). Like those criteria, they owe as much to the rationalization of the qualitative choices of naturalists as to ecology (Adams and Rose 1978). They demonstrate further the strength of the traditional nature preservation element in MAB thinking.

Perhaps for this reason, the importance given either to the potential for research or a history of research varies. In the USA for example it was seen as 'critical' (Franklin 1977), and there are other examples, such as the Mount Kulal Biosphere reserve in Kenya on which the work of the Integrated Project on Arid Lands was carried out. In fact, there was another reason for the inclusion of biosphere reserves in the MAB programme quite apart from their possible contribution to research or conservation. This was the need for a symbol of MAB in action, 'a clearly defined territorial and logistic base'. 'Thus, the introduction of the term 'biosphere reserve' was due more to an instinctive feeling that MAB's operational character and visibility in the field had to be asserted, rather than for clearly defined technical reasons' (Batisse 1982, p. 101).

The first 57 biosphere reserves were approved by the MAB Bureau in 1976. By 1977 there were 118 in 27 countries, and by 1982 there were 214 in 58 countries (Batisse 1982). The reserves are identified within the IUCN's global biogeographic classification, but selection remains in the hands of national governments. The 'success' of this Project owes much to the ease with which it could be fitted into existing moulds of action by governments; most biosphere reserves coincide with or incorporate areas already designated. Certain biomes are over-represented, particularly those in the industrialized countries such as temperate broadleaved forests (21 reserves in 1979), and highland areas (50 reserves in 1979). Environments confined to the Third World are under-represented, for example tropical humid forests (8 reserves in 1979) and savannas ((Gilbert and Christy 1981). The World Conservation Strategy (IUCN 1980) lamented that of 193 biogeographical provinces, 35 were represented in no national park or equivalent reserve, and a further 38 were inadequately covered (IUCN 1980, para 17.9).

Once again a 'global' initiative generated in the North is reduced to rather small ripples on the periphery. Nonetheless, the biosphere reserve has become a recognized and accepted element in international conservation, and although the MAB programme as a whole has failed to live up to the hopes it raised at its inception, it undoubtedly contributed through the 1970s to the growing belief that there was an ecologically sound approach to development which would be 'sustainable' and acceptable, and that this could be discovered for specific environments and circumstances through research done in new, open and inter-disciplinary ways.

The biosphere reserves represent this vision at its rosiest tint. Batisse wrote that 'experience already shows that when the populations are fully informed of the objectives of the biosphere reserve, and understand that it is in their own and their children's interests to care for its functioning, the problem of protection becomes largely solved. In this manner, the biosphere reserve becomes fully integrated - not only into the surrrounding land-use system, but also into its social, economic, and cultural, reality' (Batisse 1982, p. 107). It is quite clear that some experience at least shows this not to be the case, for example in the case of the Lufira Valley in Zaire (Schoepf 1984), described in Chapter 8. This vision - of conservation integrated with and serving some rather vaguely defined human (and hence economic) purpose - was central to the MAB Programme, and through the 1970s it was fostered by it. The identical notion resurfaces in 1980 in the sustainable development concept of the World Conservation Strategy.

Global environmentalism: the Stockholm Conference

The United Nations Conference on the Human Environment held in Stockholm in 1972 is usually identified as the key event in the emergence of global environmental concern. It was however only partly, and belatedly, concerned with the environmental and developmental problems of the emerging Third World. The primary motivation behind the UN's decision to hold an environmental conference in 1972 came from the developed world, and the initial focus was on the environmental problems of industrialization. Sweden itself was particularly concerned (as indeed it remains) about acid rain. Although the focus was (in common with so much else of 1970s environmentalism) resolutely global, it was the classic concerns of First World environmentalism, particularly pollution, which were dominant in the international forum at that time. Planning began in 1968, and a preparatory committee began meeting in 1970.

The Conference did not, however, command global support. Environmental problems and development problems had become separated, and the sense of integration and of shared problems between developing and industrialized countries was lost (Russel 1975). Neo-Malthusian ideas, whether of zero growth or lifeboat ethics, were deeply unattractive to and mistrusted by developing countries:

> some 'developing' countries felt that the concept of global resources management was an attempt to take away from them the national control of resources. Furthermore, as industrialised countries used the lion's share of resources and contributed to most of the resulting pollution, the Third World countries did not see much reason to find and pay for the solutions.
>
> (Biswas and Biswas 1984, p. 36)

A meeting of a Preparatory Committee of twenty-seven experts was arranged at Founex in Swizerland in June 1971 to iron out this divergence in view, for it seemed possible that controversy over the relative priorities to be accorded environment and development in the Third World might cause the Stockholm Conference to fail. The Founex meeting was certainly a political success, for the developing countries duly attended the Stockholm meeting. Some argue that it also succeeded in 'redefining development goals and objectives to include ecological, cultural and social factors' (Ferau 1985). Biswas and Biswas suggest that it demonstrated indissoluble links between environment and development, that 'environment and development problems are compatible, and that the false dichotomy of "environment versus development" should no longer be recognized, let alone fostered' (Biswas and Biswas 1984, p. 36). In fact the Founex meeting did not break new conceptual ground. It simply repeated the statement of faith that development and environment could be combined in some way which would optimize ecological and economic systems. The meeting was primarily aimed at allaying Third World fears about the economic effects of environmental protection policies. It promised that the Stockholm meeting would 'point the way towards the achievement of industrialization without side-effects', but it gave no idea of how this was to be achieved (Clarke and Timberlake 1982, p. 7).

It was the same at the Conference itself. Although the phrase the 'pollution of poverty' was coined, there was little discussion of the links between poverty and environmental degradation, and few of the Conference's Declaration of Principles did more than recognize the nature of the particular problems of the Third World. Even the humane and encyclopaedic popular book written for the conference, *Only One Earth* by Barbara Ward and René Dubos (1972), offers relatively little to the 'developing regions'. It recognizes the hard inheritances of colonialism and exploitative trade, and discusses the problems of population, possible policies for growth in agriculture and industry, and the question of urban environments. The synthesis, however, is global, and there is little beyond general exhortation in this volume - which became one of the classics of 1970s environmentalism - about how environment and development could be integrated in the Third World.

The Conference itself in June 1972 was attended by 113 nations. The German Democratic Republic was not invited (although the Federal Republic was), and in protest the USSR and most of the eastern European countries did not attend. Five hundred non-governmental organizations participated in a parallel 'fringe' meeting, the Environmental Forum. There were bitter debates, for example about colonialism, Vietnam, whaling and nuclear weapons testing, before compromise was reached in agreement on 26 Principles and 109 Recommendations for Action (Clarke and Timberlake 1982). The Principles were wide-ranging, from human rights (1) and disarmament (26) through to the need for environmental education and research (19 and 20). There were general exhortations about pollution (6 and 7), the need to 'safeguard' wildlife and natural resources, to 'share' non-renewable resources (5), and to cooperate over international issues. There was stress on the right of individual nations to determine population and resource policies (16 and 21).

Most importantly, in the light of the influence of the 'spirit of Stockholm', there were deliberate attempts to address the problems of the Third World. The fundamental point was that development need not be impaired by environmental protection (Principle 11). This was to be achieved by integrated development planning (13) and rational planning to resolve conflicts between environment and development (14). Furthermore, development was needed to improve the environment (8) and this would require assistance (9), particularly money to pay for environmental safeguards (12), and reasonable prices for exports (10).

Like Founex, the Stockholm meeting itself stated the need to resolve conflicts between environment and development without demonstrating how. The suggested solutions, 'rational planning' or 'integrated development', were words without substance. They owed much to the reductionist element in ecological and environmentalist thinking developing at that time. In very mild form they fitted with the 'ecofascist' vision of environmental harmony through central control (see Pepper 1984): planning was neutral and perfectable; conflicts could be planned away.

Of the 109 Recommendations for Action, only eight addressed the question of development and environment, and they were 'extraordinarily negative' (Clarke and Timberlake 1982, p. 12), concerned chiefly at minimizing possible costs of environmental protection. In the ensuing decade, there was little progress in this field, the only specific report by UNCTAD failing to identify effects of economic policies on trade, and by 1982 the debate was said to be 'largely dead' (Clarke and Timberlake 1982, p. 23).

The most conspicuous result of the Stockholm Conference was the creation of the United Nations Environment Programme (UNEP). This was established by resolution of the General Assembly of the United Nations in December 1972 to act as a Governing Council for environmental programmes, a Secretariat to focus environmental action within the whole UN system and an environment fund to finance environment programmes. UNEP was located in Nairobi in Kenya, the first UN body outside the developed world a symbolic (and politically astute) decision. The suggestion of a new full UN agency to deal with environmental problems was strongly opposed by the existing UN agencies (Ferau 1985). UNEP therefore is not a UN Agency like UNESCO or FAO, and these remain responsible for the environmental aspects of its own activities. UNEP seeks to act as a catalyst and think-tank, the 'conscience of the UN system' (Clarke and Timberlake 1982, p. 49).

Sadly UNEP's small size, poverty, relative weakness within the UN system and peripheral position in Nairobi have limited its effectiveness. It is officially a unit of the UN Secretariat, and gets administrative funds from that source, but money for projects comes from the Environment Fund. Contributions to that are voluntary, and have fallen short of needed targets. UNEP's influence on the UN Agencies has been relatively small, and they have gone about their business much as before. As Myers and Myers commented, 'we wanted an Environmental Programme of the United Nations. Instead we got an United Nations Environment Programme' (Myers and Myers 1982, p. 201).

UNEP was saddled with an impossibly broad remit, and a vague list of priorities from the Stockholm Conference. It has had several notable achievements. The Regional Seas Programme launched in 1974 has roots directly in the Stockholm Declaration on marine pollution, and represents a policy of tackling a global problem through regional action. By 1982 the programme covered ten regions and 120 coastal states (Bliss-Guest and Keckes 1982). UNEP acts as a coordinator of inter-governmental action based on an agreed Action Plan, that for the East Asian Region for example being agreed at a meeting in Manila in the Philippines in April 1981 (Bliss-Guest and Keckes 1982). Another UNEP Programme is GEMS, the Global Environment Monitoring System, begun in 1975 and centred on Nairobi (Gwynne 1982).

UNEP's most significant contribution, however, is undoubtedly its role in organizing the UN Conference on Desertification in Nairobi in 1977 (United Nations 1977). This followed a resolution of the UN General Assembly in 1974 which had discussed the Sahelian drought of 1972-4. The conference adopted a Plan of Action to Combat Desertification (PACD), which was subsequently endorsed by the UN General Assembly, including a series of recommendations on national, regional and international action. Coordination of this plan was entrusted to UNEP (Karrar 1984), although in fact other UN bodies had significant experience in the field (such as UNESCO in arid zone research).

The desertification conference was only one of a series of UN conferences broadly in the environmental field which followed Stockholm, for example the Habitat Conference on Human Settlements in Vancouver in 1976, and the UN Conference on New and Renewable Sources of Energy in Nairobi in 1981. More significant for the idea of sustainable development was the meeting of experts held at Cocoyoc in Mexico in October 1974. This achieved what Stockholm had not, and looked at environmental problems from the perspective of the Third World, and particularly the Third World poor. The resulting Cocoyoc Declaration pointed to the problem of the maldistribution of resources and to the inner limits of human needs as well as the outer limits of resource depletion. It pointed to basic needs, and called for a redefinition of development goals and global lifestyles. Obviously these ideas stemmed chiefly from the debates in development which had been emerging at the end of the First Development Decade, and were to emerge again in the global interdependence arguments of the Brandt Report (Brandt 1980). They represented the productive fusion of those debates with those of environmentalism. By the time the World Conservation Strategy emerged six years later, the two had, superficially at least, merged.

UNEP certainly played a catalytic role in the preparation of the World Conservation Strategy. The notion of a strategic approach to conservation was first considered by IUCN in 1969 at the New Delhi General Assembly, reflecting a shift in IUCN thinking towards greater concern for economic development (McCormick 1986a). Conservation and development was the theme of the 1972 General Assembly at Banff, Canada. Work on a strategy for nature conservation began in 1975, and from this restricted root the notion of a 'world conservation strategy' grew (Boardman 1981, McCormick 1986a, b). In 1975 IUCN joined UNEP, UNESCO, and FAO to form the Ecosystem Conservation Group.

In 1977 UNEP commissioned IUCN to draft a document to provide 'a global perspective on the myriad conservation problems that beset the world and a means of identifying the most effective solutions to the priority problems' (Munro 1978). Preliminary drafts were discussed at the IUCN General Assembly in Arshkhabad (USSR) in 1978. At that stage the focus was heavily orientated towards the conservation of species and special areas rather than the nature or imperatives of development (Munro 1978). Substantial changes were made to expand the focus to questions of population, resources, development and implementation (Boardman 1981). The World Conservation Strategy was eventually published in 1980. The ideas it contains are discussed in the next chapter.

3

The Ideology of Sustainable Development

The World Conservation Strategy

The World Conservation Strategy (WCS) was prepared by IUCN with finance provided by UNEP and the World Wildlife Fund. It was published in 1980 in the name of these three organizations (IUCN 1980). It had been presented to FAO and UNESCO, and publication had been delayed to include their amendments (McCormick 1986a). It was endorsed by the Ecosystem Conservation Group (IUCN 1980). In the words of the Chairman of the WWF, Sir Peter Scott, the WCS was intended to show 'how conservation can contribute to the development objectives of governments, industry and commerce, organized labour and the professions', as well as being the first time that development was suggested 'as a major means of achieving conservation, instead of being viewed as an obstruction to it' (Allen 1980, p. 7).

The WCS was the culmination of more than two decades of thinking by conservationists, particularly those in IUCN, about ways to further nature conservation on a global scale. It offered 'both an intellectual framework and practical guidance' (IUCN 1980, p. i). It was aimed at government policy-makers, conservationists and development practitioners, and was not a document to promote discussion. Instead it was intended 'to stimulate a more focused approach to the management of living resources and to provide policy guidance on how this can be carried out' (IUCN 1980, p. vi).

The WCS identifies three objectives for conservation. First, the maintenance of 'essential ecological processes' (Section 2). These are 'governed, supported or strongly moderated by ecosystems and are essential for food production, health, and other aspects of human survival and sustainable development' (para 2.1). They are called, in the non-ecological shorthand adopted, 'life-support systems', and include agricultural land and soil, forests, and coastal and freshwater ecosystems. Threats include soil erosion, pesticide resistance in insect pests, deforestation and associated sedimentation, and aquatic and littoral pollution. The second objective is the preservation of genetic diversity, both the genetic material in different varieties of locally-adapted crop plants or livestock and in wild species (Section 3). This genetic diversity is both an 'insurance' (for example against crop diseases), and an investment for the future (eg. crop breeding or pharmaceuticals) (para 3.2). The WCS's third objective is 'the sustainable development of species and ecosystems' (Section 4), particularly fisheries, wild species which are cropped, forests and timber resources and grazing land.

These objectives are then broken down into a list of priority requirements (Sections 5-7). These are drawn up on the basis of criteria of significance (how important is it?), urgency (how fast is it getting worse?), and irreversibility (Section 5). These are listed in Table 3.1. They range from the sublime intention to 'prevent the extinction of species', para 6.1) to the detailed requirements of site-protection for species conservation (para 6.8). The first two objectives, to conserve ecological processes and genetic diversity, have a relatively modest five priorities each. The first of these basically demands the rational planning and allocation of land uses: giving crops priority on the best land (but not on marginal land), and setting aside and controlling areas such as watersheds and littoral zones for appropriate management only. The conservation of genetic diversity demands site-based protection of ecosystems and the timely creation of banks of genetic material. Perhaps under the influence of the MAB Programme, a number of the familiar needs of nature conservation for protection of the habitats of rare and unique species and typical ecosystems in appropriately organized systems of reserves appear here.

The third objective, the sustainable utilization of resources, has no less than twelve priority tasks demanding linked research and action to determine and achieve resource utilization at sustainable levels. These include some which even at the time of publication were well-established, for example the regulation of international trade in wildlife products, and others such as the ingenuous suggestion to 'limit firewood consumption to sustainable levels' (para 7.10), which must surely defy policy implementation in the real world. The implications for some of this thinking are discussed below.

The WCS then discusses the priorities for national action (Sections 8-14). These are based on on the preparation of separate national strategies (by governments or NGOs) which review development objectives in the light of the conservation objectives and establish priority requirements, identify obstacles and propose cost-effective ways

Table 3.1 Priority requirements of the World Conservation Strategy

A. Ecological processes
1. Reserve good cropland for crops (para 5.1)
2. Manage cropland to high, ecologically sound standards (para 5.3)
3. Ensure that the principal management goal for watershed forests and pastures is protection of the watershed
4. Ensure that the principal management goal for coastal wetlands is the maintenance of the processes on which the fisheries depend (para 5.6)
5. Control the discharge of pollutants (para 5.8)

B. Genetic Diversity
1. Prevent the extinction of species (para 6.1)
2. Preserve as many kinds as possible of crop plants, forage plants, timber trees, livestock, animals for aquaculture, microbes and other domestic organisms and their wild relatives (para 6.4)
3. Ensure on-site preservation programmes protect
   - the wild relatives of economically valuable and other useful plants and animals and their habitats
   - the habitats of threatened and unique species
   - unique ecosystems
   - representative samples of ecosystem types (para 6.8)
4. Determine the size, distribution and management of protected areas on the basis of the needs of the ecosystems and the plant and animal communities they are intended to protect (para 6.10)
5. Coordinate national and international protected area programmes (para 6.12)

C. Sustainable Utilization
1. Determine the productive capacities of exploited species and ecosystems and ensure that utilization does not exceed those capacities (para 7.1)
2. Adopt conservation management objectives for the utilization of species and ecosystems (para 7.2)
3. Ensure that access to a resource does not exceed the resource's capacity to sustain exploitation (para 7.3)
4. Reduce excessive yields to sustainable levels (para 7.4)
5. Reduce incidental take as much as possible (para 7.5)
6. Equip subsistence communities to utilize resources sustainably (para 7.6)
7. Maintain the habitats of resource species (para 7.7)
8. Regulate international trade in wild animals and plants (para 7.8)
9. Allocate timber concessions with care and manage them to high standards (para 7.9)
10. Limit firewood consumption to sustainable levels (para 7.10)
11. Regulate the stocking of grazing lands to maintain the long term productivity of plants and animals (para 7.11)
12. Utilize indigenous wild herbivores, alone or with livestock, where domestic stock alone would degrade the environment

Source: IUCN (1980)

of overcoming them, determining priority ecosystems and species for conservation and establishing a practical plan of action. The 'strategic principles' offered are to integrate conservation and development by doing away with narrow sectoral approaches; to manage ecosystems so as to retain future options on use (this reflecting the poor state of knowledge about tropical ecosystems in particular); to mix cure and prevention, and to tackle causes as well as symptoms (para 8.6).

A number of key problems are highlighted. The first is the relative weakness of conservation in national policy-making, combined with the sectoral nature of such planning (Section 9). The proposed solution is to develop 'anticipatory and cross-sectoral' environmental policies, and to integrate conservation and development. The second problem is that environmental planning rarely allocates land uses rationally (Section 10). The solution proffered here is more and better evaluation of ecosystems, improved assessments of the environmental effects of development before implementation, and revised procedures for allocating land and water resources on the basis of ecosystem capability set against demands. The third problem is that of inadequate legislation and weak and overlapping natural resource management agencies (Section 11). The fourth is shortcomings in training and education of conservation personnel at all levels, and basic ecological data, requiring more teaching and more research, some of it internationally funded or coordinated (Section 12). The fifth is the lack of support for conservation policies, 'lack of awareness of the benefits of conservation and of its relevance to everyday concerns' (para 13.2). This failure requires more public participation in planning through community involvement, and public education. The sixth problem the WCS highlights is the lack of conservation-based rural development, 'rural development that combines short term measures to ensure human survival with long term measures to safeguard the resource base and improve the quality of life' (para 14.5).

In conclusion, the WCS turns to international actions to promote conservation (Sections 15-20), recognizing that many living resources lie partly or wholly outside national boundaries. These sections cover the obvious areas of international law and conventions, the responsibilities of bilateral and multilateral aid donors (Section 15), the cooperative management of the global commons of the open ocean, the atmosphere and the Antarctic (Section 18), and international river basins and seas (Section 19). More surprising, perhaps, is a rather unfocused call for 'international action' to conserve tropical forests and to support the Plan of Action on Desertification (Section 16), and a call for a 'global programme for the protection of genetic resource areas' (Section 17). This programme would involve the site-based protection of concentrations of economic or useful varieties (for example wild relatives of cultivars), concentrations of threatened species, ecosystems of 'exceptional diversity', and ecosystems that are poorly represented in existing protected areas. This requires financing internationally, because 'many countries particularly rich in genetic resources are developing ones that can ill-afford to bear alone the burden of their on-site protection' (para 17.11), or possibly through commercial participation or sponsorship .

Sustainable development in the World Conservation Strategy

IUCN took on the task of promoting the implementation of the WCS through national strategies and international action (IUCN 1980, para 20.7). The promotion of sustainable development formed one of IUCN's seven Programme Areas for the period 1985-7. The original plan to revise the WCS every three years gave way to progressive adaptation as national conservation strategies were produced under IUCN guidance (IUCN 1984a, McCormick 1986a, b). A number of countries in both First and Third Worlds have produced national strategies (UK Programme Organising Committee 1983, IUCN 1984b, Nelson 1987, Bass 1988). In terms of the level of nominal adoption, the WCS is therefore a success, and certainly the proliferation of the phrase 'sustainable development' owes much to the successful capture by the WCS both of media attention and of a place in development terminology.

However, opinion about the WCS is divided. Thus on the one hand, Caldwell describes it as 'the nearest approach yet to a comprehensive action-orientated programme for political change' (Caldwell 1984, p. 306), while on the other hand Redclift argues 'despite its diagnostic value the World Conservation Strategy does not even begin to examine the social and political changes that would be necessary to meet conservation goals' (Redclift 1984, p. 50). The WCS is unmistakably the child of 1970s environmentalism, and more generally it bears the hallmarks of much earlier conservation thinking. Furthermore, within it lie contradictions that reflect the shallowness of the conversion of the conservation camp after 1972 to any really new way of thinking.

The WCS bears the clear imprint of the neo-Malthusian concerns of the ecological roots of the 'new environmentalism'. The need for new approaches to resource management is underpinned by the impact of population growth and rising demand: 'the escalating needs of soaring numbers have often driven people to take a short-sighted approach when exploiting natural resources' (IUCN 1980, p. i). The neo-Malthusian message is clear, but it is tempered and moderated. Thus although it is argued that every country should have a 'conscious population policy' to achieve 'a balance between numbers and environment' (IUCN 1980, para 20.2) , this need does not receive the prominence which much environmentalist writing of the previous two decades would have given it.

The WCS represents environmentalism carefully repackaged for a new audience, but its message is essentially the same. It is a determinist vision which the WCS puts forward. Ecology (and environmental systems more generally) are presented as setting limits on human action. Notions from wildlife management are applied freely to human populations on a global scale: 'Human beings, in their quest for economic development and enjoyment of the riches of nature, must come to terms with the reality of resource limitation and the carrying capacities of ecosystems' (IUCN 1980, p. i). The WCS argues not only that ecology should determine human action in the way development is attempted, but that it does so anyway. The WCS offers a 'conservation or disaster' scenario not far removed from the classic polemics of the 'ecodoomsters' (Hill 1986).

The WCS, like IBP and MAB before it, is an attempt to find a global context for environmental action. Its concept and formulation are in part a legacy of the globalism of environmental thinking in the 1970s. One of two features which are said to 'characterize our time' is the 'global interrelatedness of actions, with its corollary of global responsibility. This in turn gives rise to the need for global strategies both for development and for conservation of nature and natural resources' (IUCN 1980, p. i). The WCS therefore sets out a global framework for conservation, although it is one which (following UNEP's model with the Regional Seas Programme) is expected to be taken on and implemented at regional and national level. Indeed, like UNEP, IUCN has an international structure yet lacks the remit and resources of the UN Agencies. The WCS is of considerable importance to these organizations. As well as being an attempt to repackage and redefine conservation ideas and to get certain ideas taken seriously, the WCS represents an attempt to find a role (and the resources to go with it) for its sponsoring agencies.

The WCS also echoes the ethics and morality of 1970s environmentalism, both scientific utilitarianism and romantic holist or vitalist thinking which was welded with aspects of scientific ecology into forms of 'bioethics' (O'Riordan 1981, Pepper 1984, Worster 1985). Wild species are to be conserved for two reasons. First, because they are useful, and second because it is in some way 'right' to do so. The former approach, which adheres to the WCS's simple conservative principle of keeping options open, is widely reflected in the work of scientific conservationists on problems such as game conservation in America (Worster 1985), or indeed Africa. The utilitarian argument of the WCS is simple: 'we cannot predict what species may become useful to us' (IUCN 1980, para 3.2). This principle underlies the relatively numerous and well thought-out proposals concerning the sustainable utilization of resources (for fisheries management, game farming, regulation of wildlife trade; see Table 3.1), which contrast markedly with the more open-ended and speculative requirements elsewhere in the document.

However, the WCS is happy to have the best of both worlds, and not to put all its eggs in the basket of utilitarianism. It does not shrink from stating moral arguments for conservation which make assumptions about the rights of other organisms, and indeed about future generations of people. Thus there is 'an ethical imperative', expressed in the belief that 'we have not inherited the earth from our parents, we have borrowed it from our children' (IUCN 1980, para 1.5). Conservation becomes a matter of moral principal, based on the fact that 'human beings have become a major evolutionary force', able to change the biosphere radically although lacking the knowledge to control it: 'we are morally obliged - to our descendents and to other creatures - to act prudently' (para 3.3). Although it is not the central argument used, the WCS also picks up the point about indigenous values placed on wildlife, pointing out that 'For a great many people, too, wildlife is of great symbolic, ritual and cultural importance, enriching their lives emotionally and spiritually' (para 4.11).

This attempt to argue for conservation along two parallel tracks is by no means accidental. It reflects, of course, the old division between technological and ecological environmentalism (Pepper 1984), or technocentrism and ecocentrism (O'Riordan 1981), which creates such schizophrenic confusion within the conservation movement and for individual conservationists. The dualism is extremely useful. On the one hand, the utilitarian argument allows conservation to be packaged in a way which is expected to be attractive to the materialism which is seen to underlie thinking about development. On the other, moral arguments can be employed where they are more effective, for example among environmentalists in industrialized countries.

Stress is laid on the need for sustainable use of ecosystems and species to ensure their continued availability 'almost indefinitely' (IUCN 1980, para 4.1). If however it is argued that industrialized economies have substantially freed themselves from primary dependence on renewable resources, and that developing economies would like to do the same, the rational argument for sustainable utilization falters. The moral argument can be rapidly marshalled to fill the gap: 'The greater the diversity and flexibility of the economy, the less the need to utilize certain resources sustainably - but by the same token the less the excuse not to' (IUCN 1980, para 4.1). While being a pragmatic recognition of real divisions within conservation, the WCS's double-barrelled justification is both versatile and robust.

Although the WCS attempts to demonstrate the indivisibility of conservation and development, the basis of its claim to marry them lies in the way they are defined. The key concept in the WCS is of course sustainability, and indeed the whole message of the WCS turns on the question of what sustainable development and sustainability mean. The opaque nature and flexibility of these terms have been discussed in Chapter 1. Curiously, nowhere in the WCS is it categorically defined. Indeed, development and conservation are defined in such a way that their compatibility becomes inevitable. Development is presented as 'the modification of the biosphere and the application of human, financial, and living and non-living resources to satisfy human needs and improve the quality of human life' (IUCN 1980, para 1.4). Meanwhile conservation is 'the management of human use of the biosphere so that it may yield the greatest sustainable benefit to present generations while maintaining its potential to meet the needs and aspirations of future generations'.

If it is taken for granted that development ought to be 'sustainable', meaning that it must be capable of being extended indefinitely for the benefit of future generations, conservation and development are of course 'mutually dependent' as the WCS argues (IUCN 1980, para 1.10), not incompatible as they have seemed in the past. The WCS is thus able to argue that development practitioners have seen conservation as irrelevant, or anti-development, because they failed to understand about 'real' conservation. Properly understood (the WCS argues), 'real' conservation would have helped avoid ecological damage and the failure of development, a failure which demonstrates that much development is not real development at all.

Environmental modification is a natural and necessary part of development, but not all such modification will achieve 'the social and economic objectives of development' (IUCN 1980, para 1.12). The solution is to give conservation a high priority in the development process, and 'to integrate every stage of the conservation and development processes, from their initial setting of policies to their eventual implementation and operation' (para 9.1). Thus every sector (health, energy, industry) involves conservation, 'that aspect of management which ensures that utilization is sustainable' (para 1.6). This integration will end the apparent conflict between conservation and development which previously obtained.

This is the core of the WCS. It certainly represents a significant repackaging of conservation. It is less clear if it really represents anything more than that, the new principle and purpose that is claimed. To a considerable extent the WCS is just conservation dressed up in new clothes. One reason for this is that the WCS is Janus-headed, addressing the very different worlds of conservation and development at the same time. Within conservation it tries to draw together the disparate threads of the utilization and preservation of nature, demonstrating that the ideas presented are familiar and established and not radically new. This it appears to do successfully, to judge by the frequency with which environmentalists cite it. To them, it seems to make a logical and effective inroad into the hitherto unfamiliar and inexplicably destructive world of economic development while reaffirming the moral basis for conservation, and finding ways to justify the protection of land in reserves and parks. Some of the flaws in this presentation are discussed later in this book.

At the same time the WCS actually has to make that advance into the development field plausible and effective. This it tries to do by stressing the utility of species and ecosystems as resources for human subsistence and development, defending even the establishment of parks and reserves on the grounds of indirect ecological benefits (such as runoff from forested watersheds) or direct economic returns from Third World 'ecotourism' (Young 1986). On this second front the WCS has been markedly less successful. Its strongest insights are those borrowed from critiques of the inhumane, monolithic and bureaucratic nature of the development process. These 'populist' critiques stress the significance and importance of indigenous cultures (e.g. McNeely and Pitt 1987), of indigenous knowledge (Brokensha *et al.* 1980, Chambers 1983, Richards 1985, 1986), the need for local participation in development and 'development from below' (Stohr 1981).

The nature and implications of these borrowings from populist development thinking are discussed further later in this chapter. These ideas were picked up and developed in the successors to the WCS (notably the Brundtland Report, discussed later in this chapter), but they are certainly present in a preliminary form in the WCS itself: 'Conservation is entirely compatible with the growing demand for "people-centred" development, that achieves a wider distribution of benefits to whole populations (better nutrition, health, education, family welfare, fuller employment, greater income security, protection from environmental degradation); that makes fuller use of people's labour, capabilities, motivations and creativity; and that is more sensitive to cultural heritage' (IUCN 1980, para 20.6).

The presentation of the WCS is well-rounded and plausible. However, despite its more or less successful comprehension of this critique of development in practice, it fails to demonstrate understanding of the debates in development which lie behind and run through this 'populist' critique. More seriously it fails to recognize the essentially political nature of the development process. This works on two levels. First, conservation - like science - is seen to be above ideology. There is no understanding of the way in which nature and culture interact, such that views of nature are created by society. There is no apparent awareness of arguments about the social production of nature (Smith 1984).

Second, the WCS suggests that conservation can in some way bypass structures and inequalities in society. It seems to assume that 'people' can exist in some kind of vacuum, outside the influence of inequality, class or the structures of power. The goal of the WCS is stated as 'the integration of conservation and development to ensure that modifications to the planet do indeed secure the survival and wellbeing of all people' (IUCN 1980, para 1.12). This is pious, liberal and benign, but it is - inevitably - ideological. Furthermore, in the context of development and social theory, it is disastrously naïve.

Like so much of the environmentalism of the 1970s, the WCS is therefore blind to the twin worlds of policy and political economy. As a result, what it has to say about development is not particularly convincing. The reasons for this, and some of the critiques and alternatives, are examined in more detail later in this book. However, first it is necessary to see the WCS in the context of other formulations of principles of 'ecodevelopment'.

Ecodevelopment

The World Conservation Strategy is not the only attempt to imbue development with environmental ideas and principles. It is only one of a series of similar formulations, all of which to a greater or lesser extent stem from the Founex and Stockholm meetings. A number of others are commonly labelled 'ecodevelopment' (e.g. Riddell 1981, Glaeser 1984a, b). This term appears to have been coined by Maurice Strong, Secretary-General of the Stockholm Conference (Boardman 1981). It was subsequently developed and promoted by UNEP (UNEP 1978), although it has been widely taken up, for example at meetings in Belo Horizonte in Brazil in 1978 and Berlin in 1979 (the International Workshop in Ecodevelopment and Appropriate Technology and the Conference on Ecofarming and Ecodevelopment held by the International Institute for Environment and Society, Glaeser and Vyasulu 1984), and Ottawa in 1986 (the *IUCN Conference on Conservation and Development*, Svedin 1987). It was also, for example, adopted by the South Pacific Commission and the South Pacific Bureau for Economic Cooperation in a Comprehensive Environmental Management Programme (Dasmann 1980).

Behind ideas of ecodevelopment lies the same awareness of the intrinsic complexity and dynamic properties of ecosystems, and the way they respond to human intervention, which underlay the WCS: the need to understand ecosystems and ensure the 'environmental soundness' of development projects (Ambio 1979, p. 115). Ecodevelopment then seeks to 'change ecosystems in the direction of higher productivity and greater relevance to human needs, while respecting the interdependence between man and nature ' (Ambio 1979, p. 115). It 'holds promise for improving the economic wellbeing of people without impairment of the ecological systems on which they must depend for the foreseeable future' (Dasmann 1980, p. 1331).

One of the foremost proponents of ecodevelopment is Sachs (1979, 1980). His approach has been extended by Glaeser (1984a, b, 1987). Their definitions have much in common with the moral arguments behind the WCS. For Sachs, ecodevelopment is 'an approach to development aimed at harmonizing social and economic objectives with ecologically sound management, in a spirit of solidarity with future generations' (Sachs 1979, p. 113). Glaeser and Vyasulu define it as 'an alternative vision of a new form of society, squarely pitted against the dynamic of overdevelopment and underdevelopment' (Glaeser and Vyasulu 1984, p. 25). Ecodevelopment is based on 'a new symbiosis of man and earth' and 'self reliance', and it focuses clearly on the satisfaction of basic needs (Sachs 1979, p. 113). Sachs does not place his ecodevelopment within the 'zero growth' tradition, arguing that it represents 'another kind of qualitative growth' requiring, among other things, low-energy strategies, resource recycling, land use and settlement planning based on ecological principles and appropriate technology.

These versions of ecodevelopment reflect a number of concerns current in development thinking in the 1970s. These represent various elements in 'neo-populist' thinking about development from within the Western development world (Byres 1979, Kitching 1982, Bideleux 1987). Neo-populism is variously defined, but Kitching offers a broad framework embracing a range of ideas which argue for 'a pattern of development based on small-scale individual enterprise both in industry and agriculture' (Kitching 1982, p. 19). Neo-populism emerged in Russia and Eastern Europe after the First World War, and extended existing populist critiques of capitalist industrialization to include the state-socialist industrialization of the USSR after the revolution.

The origins and breadth of populist and neo-populist thinking is reviewed by Kitching (1982), and Bideleux (1985) rehearses the arguments for 'communal village-based development', considering the work of (among others) Kropotkin, Chayanov and Gandhi. Kitching then provides a critique of the application of neo-populist thinking in the developing world in Tanzania (and in the writings of Julius Nyerere), in the concept of 'urban bias' (Lipton 1977), and the 'small is beautiful' thinking of Schumacher (1973). Taking the two countries most often quoted as examples of neo-populist development policies,

Tanzania and China, Kitching concludes 'In short, populism makes good social and moral criticism, and has often produced very effective political sloganeering, but on the whole makes rather flabby economic theory' (Kitching 1982, p. 140). Despite this flabbiness (or possibly because of it), these ideas have proved attractive to those approaching concepts of sustainable development and ecodevelopment, and a number have been co-opted to strengthen the developmental credentials of thinking about sustainability.

First, there has been a focus on basic needs (O'Riordan 1988): 'Ecodevelopment refers to a process which is geared to the satisfaction of basic and essential human needs, starting with the needs of the poorest and neediest in society' (Glaeser and Vyasulu 1984, p. 25). This is a direct adoption of ideas from the wider debate about basic needs which emerged in the Second Development Decade of the 1970s (e.g. McHale and McHale 1978, Morris 1979, Streeten *et al.* 1981, Stewart 1985). Critics of this approach point out its political weakness, resulting from the strength of vested interest and power which oppose wealth redistribution and decentralization (Lee 1981, Seddon 1983). Its adoption by ecodevelopment thinkers is further evidence of their failure to look beyond morality and the principles of rational planning to address the political economy of the development process.

Second, it is argued that ecodevelopment demands participation. Sachs calls for 'participatory planning and grass-roots activation' (Sachs 1979, p. 113). Glaeser and Vyasulu see participation where 'people who are to be affected by changes which they have decided are desirable cooperate voluntarily in the process of implementing the changes by giving them direction and momentum' (Glaeser and Vyasulu 1984, p. 26). This is in itself a naïve suggestion, but it is typical of thinking about ecodevelopment which advocates non-hierachical systems of organization and government. Thus Galtung (1984) differentiates between 'alpha' social systems, which are hierachical, unlimited in size and tending towards uniformity, and beta systems, which are horizontal, limited in scale and inclined towards diversity. Ecodevelopment requires beta structures, but set within a matrix of a benign, flexible, communicating and restraining alpha system. Galtung's ideas owe little to social theory and a great deal to the diversity-stability debate in ecology in the 1960s (Margalef 1968). Nonetheless, they reflect the element of anarchism in environmentalism which provides one of the central themes of 'green' development thinking (e.g. Bookchin 1979, Roszak 1979) as well as notions of participation in planning and 'development from below' (Stohr 1981).

Third, ecodevelopment picks up the ideas of appropriate and intermediate technology. This is hardly surprising, since *Small is*

Beautiful (Schumacher 1973) was one of the definitive books of the 'environmental decade' as well as a successful critique of development. Schumacher wrote a first version of the key chapter of that book, 'Buddhist economics', in 1955 when he was economic adviser to Burma (McRobie 1981). In 1961 he spoke to an international seminar at Poona in India, and the following year wrote a report for the Indian Planning Commission, at the instigation of Nehru, developing the same line of argument. This was coolly received (McRobie 1981), but his ideas on intermediate technology received wider publicity at a seminar in Cambridge in 1964 (Robinson 1971). In May 1965 the Intermediate Technology Development Group (ITDG) was formed, its first major work being a guide to the availability of hand and small-scale tools useful in rural development.

These formulations of ecodevelopment outside the WCS obviously differ from it in the scope of comprehension of socio-economic aspects of the development process. UNEP for example emphasizes the decentralization of bureaucracy, a disaggregation of development focus ('the achievement of sustainable development at local level'), self-reliance and self-sufficiency, the priority given to meeting basic human needs, public participation and equitable distribution (Ambio 1979). The goal of ecodevelopment is 'to pursue economic development that relies for the most part on indigenous human and natural resources and that strives to satisfy the needs of the population, most of all the basic needs of the poor' (Glaeser 1984a, p. 11). Singh, writing from the perspective of a Malaysian non-governmental conservation organization, describes it as 'development of the people, for the people by the people' (Singh 1980, p. 1350), using explicitly socio-economic terminology to explain the context of an ecologically-based view of sustainable ecosystem use.

While obviously closely related in many ways, these versions of ecodevelopment offer a very different agenda from that of the WCS. It is one which avoids a number of the criticisms which can be levelled at the IUCN document, particularly its divorce from the language and ideas of economics and politics. Indeed, UNEP recognized the implications of such ideas for developing countries in terms of social and economic structure, the inevitable political implications of such changes, and the constraints of capital, technical knowledge and manpower. This perception is hardly revolutionary, but it is important because essentially environmentalist ideas are being placed in a development matrix in a way in which the WCS failed to do.

There have been a number of attempts to place ecodevelopment in a specific development context, notably by Glaeser (1984a, b). His aim is a modified strategy for development whose starting point is 'the needs of the people' (Glaeser 1984a, p. 10). It involves the use of indigenous human and natural resources and the 'long term preservation of the ecological basis of economic development' (p. 12). Such an approach he sees as a different road to development from the general scientific and technical development of agriculture, but not an alternative, and not part of 'back-to-nature' romanticism. Glaeser describes the 'agrotechnical foundation' of ecodevelopment (what he terms ecological farming or 'ecofarming') in the context of a study in the Usumbara hills in northeast Tanzania in 1976. This is an empirical study which generates a series of practical ecological innovations aimed at improving the biological productivity of agriculture. These suggestions involved agroforestry techniques, crop mixes, a certain level of tolerance of weeds, selection of pest-resistant plant varieties, and reduced mechanical tillage. This approach fits 'into sociological traditions and values and relies on existing natural and human resources' (Glaeser 1984a, p. 21). It is both appropriate to, and helps preserve, the small decentralized and non-capital-intensive farming units of the area.

Not all formulations of ecodevelopment stem, however, from an essentially libertarian position. Others take their origins from more authoritarian aspects of environmentalism. This approach is not confined to scientists with an overdeveloped view of the potential of rational centralized planning to generate human welfare, but is shared by some social scientists. Thus the ecodevelopment offered by Riddell (1981) is an economist's attempt to develop 'an alternative to growth imperative models'. Riddell defines underdevelopment (the condition of the poor Third World) and 'overdevelopment' (the industrialized countries of the First World), and defines ecodevelopment an an 'altogether different' alternative to both (Riddell 1981, p. 5). It involves self-reliance and self-constraint, 'neither to capture other countries' resources nor to give way to an interloper', a 'best fit' attempt to 'optimise the balance between population numbers, locally available resources and culturally desired lifestyles' (p. 5). It is not a hopeful vision, more or less writing off the possibility of 'development' to living standards of the North in the Third World. Riddell argues that 'it is better for the governments of low income nations to contemplate the realities of human progress in preference to the illusion of economic growth. Whereas economic growth for Southern nations is desirable but illusive, social improvement is very possible' (p. 158).
Riddell cites Burma, China and Tanzania as contrasting examples which 'approach the ecodevelopment ideal'. These are, of course, the countries held up as examples by neo-populist thinkers in development (Kitching 1982), and it is as an offshoot of this thinking that Riddell's ideas are best seen. However, they also owe much to the globalist perspective of 1970s environmentalism and 'interdependence' development thinking (Brookfield 1975, Brandt 1980, 1983, Corbridge 1986). The importance of ideas of interdependence to sustainable development thinking is discussed in the last section of this chapter.

Table 3.2 Principles of ecodevelopment

1. Establish an ideological commitment
2. Sharpen political and administrative integrity
3. Attain international parity
4. Alleviate poverty-hunger
5. Eradicate disease-misery
6. Reduce arms
7. Move closer to self-sufficiency
8. Clean up urban squalor
9. Balance human numbers with resources
10. Conserve resources
11. Protect the environment

Source: Riddell 1981

Riddell calls for 'downward adjustments' in the level of material consumption in Northern nations to allow for 'upgrading' in the South. This does not immediately suggest itself as a realistic or practicable suggestion, but Riddell is undaunted and goes on to offer a package of actions which could bring it about (see Table 3.2).

These proposals also hardly invite serious attempts at implementation. They resemble most of all the idealist, technocratic and 'apolitical' ideas of 'global management' environmentalist writers discussed in Chapter 2. Riddell's ecodevelopment demands a tightly controlled society with 'political cohesion' to 'generate the necessary political will to break with Northern insistence to compete on Northern terms', the 'scotching' of 'entrenched corruption and petty inefficiencies' in bureaucracy and a 'soundly structured' society within which birth control through 'socio-economic measures' can be implemented (Riddell 1981). The sustainable utilization of resources and the realization of basic needs is to be achieved through a strongly authoritarian state.
The social and political content of this agenda, similar to that in some of the neo-Malthusian writing of 1970s environmentalism, is addressed in the next chapter in the context of radical responses to 'green' thinking about environment and society.

The Brundtland Report

Although the World Conservation Strategy and other approaches to ecodevelopment differ in the extent to which they address social and economic problems, both suffer from being primarily theoretical rather than applied, and concerned with the local scale of development rather than the global. These theories are good in laying out principles for peasant agriculture, or expounding the benefits of appropriate technology and the importance of indigenous value systems and technical knowledge. They barely begin to address the larger issues of national economic management (questions for example of the relative weight given to different sectors, or the pros and cons of economic closure), let alone the burning questions of international political economy. They say nothing beyond bland generalities about the 'North-South divide', or interdependence, and as we shall see later nothing at all about the various radical theories on the global economy. The global scope of 'ecodevelopment' analysis, borrowed still from 1970s neo-Malthusianism, embraces neither the real world and the practical politics of international development, nor the theoretical ideas being discussed in development studies. For this reason, while ecodevelopment was acceptable precisely because of its blandness, it remained marginal to debates in development.

However, ecodevelopment has moved in two directions which go some way to remedying this deficiency. First, some writers have sought to make explicit the links between ecodevelopment ideas and development theory. A body of theory has developed which is closely allied to the sustainable development or ecodevelopment literature yet which is integrated with radical development theory. This 'red-green' thinking is analysed in Chapter 4. Second, ecodevelopment has been brought into the established political arena of international development through the establishment in December 1983 of the World Commission on Environment and Development at the call of the UN General Assembly. In 1987 its report, *Our Common Future* (Brundtland 1987), was presented to the UN General Assembly.

Our Common Future claimed a very specific heritage. The Chairman wrote 'After Brandt's *Programme for Survival* and *Common Crisis*, and after Palme's *Common Security*, would come *Common Future*' (Brundtland 1987, p. x). Like these predecessors, its target was multilateralism and the interdependence of nations: 'the challenge of finding sustainable development paths ought to provide the impetus - indeed the imperative - for a renewed search for multilateral solutions and a restructured international economic system of cooperation' (Brundtland 1987, p. x). This concern with multilateralism is blended with the globalist concerns of environmentalism to present a picture which verges on the sugary. Thus the aim of the sustainable development strategy outlined by *Our Common Future* is 'to promote harmony among human beings and between humanity and nature' (Brundtland 1987, p. 65).

The Brundtland Commission is important for several reasons. First, it obviously and rather self-consciously attempts to recapture the 'spirit of Stockholm 1972' which was so celebrated by environmentalists in the early 1970s and whose demise as recession bit was so lamented. The World Conservation strategy tried to do exactly the same thing, of course, but Brundtland achieves this resuscitation far more expertly and effectively, largely because of its origins in the UN General Assembly and not out in the wildwoods of UNEP and IUCN.

Second, *Our Common Future* places elements of the sustainable development debate within the economic and political context of international development. Its starting point was deliberately broad, and a move to limit its concern simply to the 'environment' was firmly resisted:

> This would have been a grave mistake. The environment does not exist as a sphere separate from human actions, ambitions, and needs, and attempts to defend it in isolation from human concerns have given the very word 'environment' a connotation of naivety in some political circles.
>
> (Brundtland 1987, p. xi)

Third, Brundtland placed environmental issues firmly on the political agenda. Arguably, of course, it simply reflected the de facto situation created by its predecessors, but nonetheless it achieved something that Stockholm, UNCOD and the WCS failed to do, and got the UN General Assembly to discuss environment and development as one single issue.

The Brundtland Report therefore starts from the premise that development and environmental issues cannot be separated: 'It is therefore futile to attempt to deal with environmental problems without a broader perspective that encompasses the factors underlying world poverty and international inequality' (Brundtland 1987, p. 3). This view is at the same time less abstract and less simplistic than that of the WCS. The argument is not presented in the rather general terms of linkages between sustainable use of ecosystems and human wealth and welfare, nor is the spectre of 'ecodisaster' prominent. Instead, the essentially reciprocal links between development and environment are drawn more explicitly: 'many forms of development erode the environmental resources on which they must be based, and environmental degradation can undermine economic development'. Furthermore, the links, again reciprocal, between poverty and environment are also recognized, poverty being seen 'as a major cause and effect of global environmental problems' (Brundtland 1987, p. 3).

In *Our Common Future* the definition of sustainable development ('development that meets the needs of the present without compromising the ability of future generations to meet their own needs', Brundtland 1987, p. 43) is based on two concepts. The first is the concept of basic needs and the corollary of the primacy of development action for the poor. The second involves the idea of environmental limits. These limits are not, however, those set by the environment itself, but by technology and social organization. This involves a subtle but extremely important transformation of the ecologically-based concept of sustainable development, by leading beyond concepts of physical sustainability to the socio-economic context of development. Physical sustainability could be pursued in 'a rigid social and political setting', but it cannot be secured without policies which actively consider issues such as access to resources and the distribution of costs and benefits. In other words the sustainable development of *Our Common Future* is defined by the achievement of certain social and economic objectives and not by some notional measurement of the 'health' of the environment. Whereas the WCS started from the premise of the need to conserve ecosystems and sought to demonstrate why this made good economic sense (and - although the point was underplayed - could promote equity), *Our Common Future* starts with people, and goes on to discuss what kind of environmental policies are required to achieve certain socio-economic goals. The answers (perhaps unsurprisingly) are remarkably similar. The premisses, however, do differ, and of the two *Our Common Future* is by far the more effective document.

The elements of the sustainable development ideas in *Our Common Future* are listed in Table 3.3. They represent an interesting blend of environmental and developmental concerns. Among the former is the need to achieve a sustainable level of population. This is a softened version of the neo-Malthusian message which recognizes the greater demands on resources made by a First World child, but still arguing that population should 'be stabilised at a level consistent with the productive capacity of the ecosystem' (Brundtland 1987, p. 56). Other rather familiar concerns are conserving (and enhancing) the resource base, and reorientating technology, particularly with regard to risk. Prominent among the latter is the fundamental concern with meeting basic needs. Prominent too is the need to 'merge' environment and economics in decision making. Most prominent of all, however, is the focus on growth; economic growth is seen as the only way to tackle poverty, and hence to achieve environment-development objectives. It must, however, be a new form of growth, sustainable, environmentally aware, egalitarian, integrating economic and social development:

'material- and energy-intensive and more equitable in its impact' (Brundtland 1987, p. 52). Above all, the Brundtland Report's vision of sustainable development is predicated on the need to maintain and revitalize the world economy. This means 'more rapid economic growth in both industrial and developing countries, freer market access for the products of developing countries, lower interest rates, greater technology transfer, and significantly larger capital flows, both concessional and commercial (Brundtland 1987, p. 89).

Table 3.3 Critical objectives for environment and development policies that follow from the concept of sustainable development

1. Reviving growth
2. Changing the quality of growth
3. Meeting essential needs for jobs, food, energy, water and sanitation
4. Ensuring a sustainable level of population
5. Conserving and enhancing the resource base
6. Reorientating technology and managing risk
7. Merging environment and economics in decision making

Source: Brundtland 1987, p. 49

However, this prescription is based on an economic and not an environmentalist vision. It uses some of the language of 1970s environmentalism, but not the questioning of growth or technology: in the classification of Cotgrove (1982) it is firmly cornucopian rather than catastrophist. It is poverty which puts pressure on the Third World environment, and it is economic growth which will remove that pressure. Furthermore it is only the ending of dependence which will enable these countries to 'outpace' their environmental problems. But what of the pressures of that growth itself? What about demands for energy and raw materials, about pollution? *Our Common Future* hopes to have its cake and eat it: 'The Commission's overall assessment is that the international economy must speed up world growth while respecting environmental constraints' (Brundtland 1987, p. 89). It does not say how this balancing trick is to be achieved. While more realistic about the development context of the sustainable development debate, *Our Common Future* moves away from the concerns of its predecessors, and does not offer new insights in this area. What does this form of sustainable development require? Quite simply a restructuring of national politics, economics, bureaucracy, social systems, systems of production and technologies, and a new system of international trade and finance (Table 3.4).

Table 3.4 Requirements of a strategy for sustainable development

1. A political system that secures effective citizen participation in decision making
2. An economic system that is able to generate surpluses and technical knowledge on a self-reliant and self-sustained basis
3. A social system that provides for solutions for the tensions arising from disharmonious development
4. A production system that respects the obligation to preserve the ecological basis for development
5. A technological system that can search continuously for new solutions
6. An international system that fosters sustainable patterns of trade and finance
7. An administrative system that is flexible and has the capacity for self-correction

Source: Brundtland 1987, p. 65

This is no small agenda. Sustainable development must be global in scope and internationalist in formulation. This demands first that 'the sustainability of ecosystems on which the global economy depends must be guaranteed', and second equitable exchange between nations. It is this latter requirement which lifts the Brundtland Report out of the mould of previous ecodevelopment writing and places it alongside the global interdependence arguments of Brandt.

This represents what is probably the most significant feature of *Our Common Future*. Environmentalists see it as a logical extension of the World Conservation Strategy, and welcome its apparently new and authoritative tone and credibility in the way it handles development terminology. In fact it is only rather secondarily the fruit of environmentalism. It is primarily an extension of the thinking of the Brandt Reports *North-South* and *Common Crisis* (Brandt 1980, 1983). Indeed, while the WCS represented the attempt by conservationists to capture the rhetoric of development to repackage old ideas, *Our Common Future* is the result of the reverse process. The existence of a global environmental crisis is cited as evidence for the need for a multilateralist solution: 'At first sight, the introduction of an environmental dimension further complicates the search for cooperation and dialogue. But it also injects an additional element of mutual self-interest, since a failure to address the interaction between resource depletion and rising poverty will accelerate global deterioration' (Brundtland 1987, p. 90).

Interdependence and sustainability

In *Our Common Future*, the establishment of the sustainable use of resources is shown to demand more than simply dealing with the micro-scale questions of industrial methodology or the issues of refining or reforming project planning procedures. It needs to embrace international trade and international capital flows. Patterns of international trade (for example in hardwood timber products) that impoverish Third World countries also promote unsustainable resource use. Lack of external capital (as aid and rescheduling of debts) limits the improvement of living standards in Third World countries: 'Without reasonable flows, the prospect for any improvements in living standards is bleak. As a result the poor will be forced to over-use the environment to ensure their own survival' (Brundtland 1987, p. 68).

Debt, poverty and population growth restrict the capacity of developing countries to adopt environmentally sound policies. The proposed solutions involve: first, increased capital flow to the Third World, but redirected to promote 'sustainable' projects; second, new deals on commodity trade (particularly attention to hidden pollution costs of industrial processes in developing countries); and third, ending protectionism and reforming transnational investment to ensure 'responsibility' (for example through technology transfer and environmentally sound technologies) (Brundtland 1987, Chapter 3). This analysis of the potential for productive reform of world trade and finance stems from a very particular vision of the working of the world economy and the pattern of economic and political forces. *Our Common Future* is, as its author hoped, the close successor to the Brandt Reports, based on multilateralism, and notions of global cooperation and dialogue.

The Brandt Reports set out much of the ground on which Brundtland builds, both in terms of general principles and in the approach to (if not the priority given to) environmental problems. *North-South* (Brandt 1980) discusses measures 'which together would offer new horizons for international relations, the world economy, and for developing countries' (Brandt 1980, p. 64). Those 'new horizons' include the environment, both globally ('the biosphere is our common heritage and must be preserved by cooperation', p. 73), and in the countries of the South themselves. The package argued for by the Brandt Commission involved growth (in both North and South), 'massive transfers' of capital, expansion of world trade, the end of protectionism, an orderly monetary system, and a move towards international equality and peace. Like Brundtland, Brandt argues that only the abolition of poverty will bring an end to population growth, and again that this is a global problem, and not one confined to the Third World alone: 'The conquest of poverty and the promotion of sustainable growth are matters not just of the survival of the poor, but everyone' (p. 75). Like other elements in the Brandt proposals, poverty requires multilateral action, not just because of the obvious moral imperative to eradicate poverty, but because of mutual self-interest. Other imperatives 'rooted in the hard self-interest of all countries and people' reinforce the claim of human solidarity (p. 77).

The Brandt Reports are themselves part of a longer evolution of thinking about economic interdependence (cf. Brookfield 1975). This essentially began with the Bretton Woods system of international financial management (the World Bank family, initially the International Monetary Fund and International Bank for Reconstruction and Development) in 1944. This was based on an essentially Keynesian vision, to create a stable, growing and interdependent world economy, 'an environment for liberal trade and to promote economic cooperation' (World Bank 1981). In the 1950s and 1960s the world economy, and international trade, indeed grew. However, fundamental problems were emerging by the 1960s, and in 1971 fixed exchange rates were abandoned, leading to volatility in currency markets, and eventually the destabilization of oil prices and the oil crisis of 1973 (Strange 1986).

The oil crisis coincided with the sudden flowering of concern about 'limits to growth'. This too stressed global interdependence, but with an altogether different message about the desirability and possibility of continued growth. The problem of the attitude of such 'global' environmentalism to aid and economic growth in the Third World at the time of the Stockholm Conference in 1972 has been discussed in Chapter 2. The reconciliation of these difficulties created formal statements of sustainable development. It is not surprising that those ideas should be based on the resurgence of Keynesian thinking represented by the Brandt Reports, a return to principles of an organized, managed and growing world economy.

The approach taken by Brandt and Brundtland is just one position among many on the world economy, and it has many critics (e.g. Seers 1980, Frank 1980, Corbridge 1982, 1986). *North-South* argues that the protectionist trade policies of industrialized countries are the root cause of global economic problems and particularly persistent slow growth in the South. Tariff barriers and quotas stifle Southern economies and cause stagnation of Northern economies as Southern markets shrink. The solution is to open up the world economy, to pump capital and technical aid into the South to encourage trade, and to accept economic restructuring in the North (Brandt 1980, p. 186).

However, this analysis fails to demonstrate adequately the alleged dependence of the North on Southern markets. Furthermore, on the scale of individual countries, the effect of trade liberalization is at best difficult to predict and at worst deleterious (Corbridge 1982, 1986). *North-South* also adopts a caricature of development in the South and the constraints upon it. It is assumed that actions on the international scale will actually reach and benefit the poor, thus ignoring the problem of political and economic structures within Third World countries (cf. Frank 1980, Seers 1980). Similarly, *North-South* assumes that political and economic interests in the North are uniform, explicitly articulated and susceptible to rational debate. *North-South* is 'a text explicitly written for Northern politicians and appealing (apparently) to their instincts for enlightened self-interest' (Corbridge 1986, p. 222). They are expected to recognize the mutual benefits from global economic cooperation and change their policies to achieve it, but Corbridge argues that they cannot do this because the constraints of political economic structures are too great. Certainly no such response has occurred; the first Brandt initiative foundered in the early 1980s amid growing protectionism, and the summit at Cancun in Mexico in 1981 achieved no breakthrough. In development (as in arms control), *North-South* adopts an unrealistic picture of the power and logic of capitalism, offering only a 'tepid programme of political action' (Corbridge 1982, p. 263) to achieve ambitious ends.

Development cannot be achieved by tinkering with world trade, but only by altering the relations of production within Third World countries and globally. Brandt avoids these issues, and as a result his Commission's concept of mutuality is 'too imprecise and too impotent to serve as a useful guide to political action on behalf of the world's poor' (Corbridge 1982, p. 253). Certainly the world has become more interconnected economically and financially. Indeed, Strange (1986) describes it as a 'casino of capitalism' with accelerating flows of money and rising uncertainty. This is an interdependence of transnational investment, banking and production manufacturing, very different from Brandt's calm managerial picture of an interdependent world economy.

The 'mutuality' of Brandt and Brundtland is therefore naïve. *Our Common Future* shows greater awareness of the real world of economics and politics than its predecessors like the WCS, but it comes no closer to explaining how that system works. Redclift (1987), writing before the publication of *Our Common Future*, suggests that the Brundtland Commission would produce a radical departure from previous writing on sustainable development. This has not proved to be the case. Certainly, *Our Common Future* sets those ideas in a more credible matrix of development thinking, but it places it firmly in only one of the camps of development thinking, in a far from commanding position. There is a great gulf between the rather comfortable Keynesian reformism of *North-South* and *Our Common Future* and other more radical and 'greener' models of development. These are the subject of the next chapter.

4

Green Development?

Greens, growth and industrialization

It is not entirely surprising to find a wide diversity of ideas marshalled under the title of sustainable development, since both its intellectual origins and the movement from which it emerged are themselves so diverse. This diversity gives great strength, since it makes it extremely difficult for protagonists to find a clear target. It stems from two characteristics. First, sustainable development is intensely synthetic. Part of its attraction is the way in which it pulls together diverse ideas and blends them, often uncritically, into an apparently seamless whole. Propagandists tend to claim a unity for sustainable development thinking which is misplaced; its true strength is its robust and polyglot heterogeneity. Sustainable development's second characteristic is the apparent ease with which different ideas about development are grafted on. It is in this way that sustainable development has become progressively more sophisticated with regard to its treatment of social and economic aspects of development, without most of its proponents ever explicitly discussing social theory or political economy as such.

A number of the elements in these approaches to sustainable development or ecodevelopment are commonly labelled 'green', in the sense of being related to 'alternative' or 'green' political ideas (Capra and Spretnak 1984). However, the 'green' label has become so widely used in recent years that its currency has become seriously debased. It is used loosely to refer to a great many ideas of diverse and often conflicting origin. Those discussed so far are reformist, and are constructed within the context of continued capitalist growth. They focus on the potential for fairly minor reforms of the existing economic system involving new approaches (for example rational planning of land use and ecosystem exploitation, people-orientated and 'bottom- up' development planning). Thus *Our Common Future* attempts to identify economic growth that yields a new broader distribution of economic goods and avoids environmental damage.

This cooption from development thinking of neo-populism and Brandt-style mutuality/multilateralism means that sustainable development is firmly anchored within the existing economic paradigms of the industrialized North. This might be called the approach of 'green growth', and it is directly parallel to the so-called 'green capitalism' and 'green consumerism' which have become features of retail marketing in Europe the 1980s . This kind of 'greening' owes more to labelling and marketing than it does to any real sea-change in ideology or the organization of production. The use of the word 'green' to describe it belies the existence of other essentially radical critiques of political economy. While it is futile to argue whether these better deserve the label of 'green' ideas, they certainly claim it. Moreover they are quite distinct from the reformist capitalism of the environmentalist critique in much sustainable development writing. These more radical ideas are the subject of this chapter.

These 'radical green' views are by no means internally or mutually consistent. Perhaps unsurprisingly, they share with their un-radical cousins in sustainable development thinking a certain chaotic diversity of thought. Some are 'radical' chiefly in the sense that they launch more robust and forthright attacks on accepted patterns of development. Others claim strong links with radical thought in anarchist or Marxist traditions. Along with the issue-based campaigning of First World environmental groups they represent what Chris Patten, then British Minister for Overseas Development, described in January 1987 as 'eco-hysteria', which he saw as 'a country cousin of the gloomy predictions that the poor of the world can only be helped if the world itself is totally transformed' (Patten 1987). These radical green critiques of development certainly do raise questions of the basic nature of the world economy. Most do not fit comfortably into conventional patterns of capitalist development planning and development aid.

One critique of development which most environmentalists embrace is that which questions the viability of continued economic growth. The work of the 'zero growth' school (Mishan 1969, 1977, Daly 1973, 1977), whom Corbridge calls the 'physicalists' (Corbridge 1986, p. 194), is a familiar element in environmentalist critiques of economics. It is radical largely in the literal sense that it demands fundamental change in economic systems. The rise of the 'futures' debate (Cole 1978) as a part of 1960s and 1970s environmentalism has been described in Chapter 2. 'Limits to growth' ideas have occasioned considerable hostility from economists (e.g. Beckerman 1974, Rostow 1978, Kahn 1979, Simon 1981). Beckerman, for example, states the orthodox position: 'a failure to maintain economic growth means continued poverty, deprivation, disease, squalor, degradation and slavery to soul-destroying toil for countless millions of the world's population' (Beckerman 1974, p. 9).

In some forms, 'zero growth' ideas are both naïve and trivial, and are easily caricatured and dismissed. Other formulations are better argued, notably perhaps Daly's idea of 'steady state economics' (Daly 1977). Daly's starting point is his 'impossibility theorem': that a high mass-consumption economy on the US style is impossible (at least for anything other than a short period) in a world of 4 billion people. He points out that the rich countries of the world have wanted to place the burden of scarcity on the poor countries through population control, while the poor want it borne by the rich through limiting consumption. Both want to pass as much as possible of the burden onto the future. The simple (but as he points out, hopelessly utopian) solution is for the rich to cut consumption, the poor to cut population and raise consumption to the same new reduced 'rich' level and both to move towards a steady state at a common level of capital stock per person and stabilized or reduced population. This idea fails because of the lack of international goodwill, and internal class conflicts within Third World societies. Describing the case of Brazil in some detail, he says:

> Now that the Brazilians have learned to beat us at our own game of growthmanship, it seems rather ungracious to declare that game obsolete. We can sympathize with Brazilian disbelief and suspicion regarding the motives of the neo-Malthusians. But the dialectic of change has no rule against irony.
>
> (Daly 1977, p. 166)

Daly is a distinguished exception to the general rule that 'zero growth' ideas have rarely been developed in a broad (or sound) enough way that they embrace the Third World and address the question of the world economy. In some senses this is unsurprising, since they are part of an environmentalism which is itself a phenomenon of the First World. It is widely argued that recent environmentalism is the fruit of societies, and groups in society, enjoying material prosperity and yet holding post-materialist values (see for example Cotgrove and Duff 1980). Such groups, the new 'middle class radicals' (Cotgrove and Duff 1980) have in general not been concerned with the extension of debates about growth to address the question of the global economy, and the relation between growth and poverty on the global periphery. It is notable that the radical economic restructuring required by zero growth has found no place in the mainstream works on sustainable development discussed in Chapter 3, the *World Conservation Strategy* and *Our Common Future*.

Green development in a capitalist world

There are, however, 'green' critiques of development which do embrace political economy. One of the most important is the feminist analysis of the role of patriarchy in capitalist accumulation, and its dual subjugation of both women and nature (Mies 1986, Shiva 1988). One element in this subjugation is the rise of modern scientific worldviews. Shiva sees this as a 'masculine and patriarchal project which necessarily entailed the subjugation of both nature and women' (Shiva 1988, p.15). Shiva's analysis builds on an account of the struggles of women for the 'protection of nature', particularly in India, describing for example womens' active roles in the 'chipko' movement, in social forestry and issues of the commons. She argues that a gender-based ideology of patriarchy underlies ecological destruction, and she calls for a 'non gender-based ideology of liberation' (Shiva 1988, p. xvii).

Thus in the forests of India, reductionist science, driven by capitalist profit maximization, has marginalized women and degraded their environment. The destruction of the forest and the displacement of women are both structurally linked to the reductionist (and capitalist) paradigm of science. Attempts to deal with deforestation, stemming from the same patriarchal science and capitalism, exacerbate both the crisis of human survival and that of environmental degradation. Salvation, the recovery of 'the feminist principle in nature', and of the view of the earth as sustainer and provider, is to be found in the capacity of women, specifically Third World women, to challenge established male-dominated modes of thinking about knowledge, wealth and value. Shiva argues that 'the intellectual heritage for survival lies with those who are experts in survival' (Shiva 1988, p. 224). Only they have the knowledge and experience 'to extricate us from the ecological cul-de-sac that the western masculine mind has manoeuvred us into' (p. 224).

Not all 'green' critiques of capitalism are feminist. Friberg and Hettne (1985) attack both capitalism and Marxism, offering a counterpoint to both 'blue' (market, liberal, capitalist) and 'red' (state, socialist) development strategies, and trying to set out a 'green alternative' in world development. This 'Green Counterpoint' opposes the institutionalization of what they call the 'modern complex' in 'the bureaucracy, the industrial system, the urban system, the market system, the techno-scientific system, and the military-industrial complex' (Friberg and Hettne 1985, p. 207). This green position embraces elements of romanticism, anarchism and utopian socialism, but they suggest it must not be interpreted 'simply as nostalgic conservatism' (p. 207). They suggest that 'the clearest manifestation of the Counterpoint' is probably populism, but Marxism too 'is broad and complex enough to contain its own mainstream-counterpoint contradiction' (p. 207). Modern 'Green' ideas differ from neo-populism in containing elements of 'an ecological consciousness (encompassing the total global ecological system)', and a 'strong commitment to a just world order' (p. 208).

Friberg and Hettne write within the context of a critique of the capitalist world system which is essentially Marxist (Wallerstein (1974). Radical development theory is widely discussed, for example by Chilcote (1984), Forbes (1984) and Corbridge (1986), and is not reviewed here. Friberg and Hettne's perspective shares a realization of the need to transform the world system if development (the achievement of 'just equitable and therefore humane' conditions of life for all the world's inhabitants) is to be achieved: 'it is only through a full confrontation with the logic and dynamics of the world-system that alternative and realistic development policies and perspectives can be conceived' (Addo *et al.* 1985, p. 2). Development cannot be achieved within the capitalist world-system:

> The bankruptcy of dominant development models, the deterioration of living conditions virtually everywhere, the sharpening of conflicts within and between nations, and the destruction of the foundations of existence should overwhelm the illusion held for so long of the possibilities of developmental transformation within the capitalist world-system.
>
> (Addo *et al.* 1985, p. 2)

Development is therefore necessarily social transformation, and is 'anti-systemic' in the sense that it must be directed at the damaging and crisis-creating features which are an integral part of the world-system.

The 'Green Counterpoint' is 'opposed and dialectically related' to the dominant development paradigm (Friberg and Hettne 1985, p. 207) which Aseniero calls 'developmentalism'. This is the philosophy of the modern world-system: 'the processes of modernization, economic growth and nation-state building, posited now as the development goals of the underdeveloped countries, are historically the constitutive processes of the modern world-system' (Aseniero 1985, p. 51). Developmentalism underpins conflicting theories: 'progress, growth, development - through these metaphors runs a basic theme that has its precise date of emergence in the history of thought' (p. 55).

Friberg and Hettne argue that this dominant development paradigm embraces both capitalist and socialist approaches to development: 'the capitalist societies of the West and the state socialist societies of the East are two varieties of a common corporate industrial culture based on the values of competitive individualism, rationality, growth, efficiency, specialization, centralization and big scale' (Friberg and Hettne 1985, p. 231). There is therefore a continuum between two opposing poles of the state ('red') and the market ('blue'). To these, green thinking offers a total counterpoint.

Friberg and Hettne argue that this development paradigm is dominated by an evolutionary perspective. This implies that development is directional and cumulative, that it is predetermined and irreversible, and that it is necessarily progressive. They challenge all these assumptions, arguing instead that it is the development process itself which is at fault. They therefore dismiss dissimilar theoretical approaches on the same grounds, for example the view of development as modernization and the 'stage' theories of development are rejected along with the whole field of *dependencia*' or 'dependency theory' (notably Frank 1966, but including many others; Lehmann 1986) as well as 'world-systems' theory (Wallerstein 1974).

Friberg and Hettne reject the evolutionary notion of progress and the idea that development is a predetermined and irreversible process. People make development, and 'human consciousness and will' are decisive elements within the process (Friberg and Hettne 1985, p. 215). They argue that industrial society is just as utopian as the 'counter utopias' erected against it (see for example Reich 1970). Industrialism is a partially-realized utopia with its roots in science and backed by power. The green attack on modern society mounted by Friberg and Hettne is essentially cultural and not economic. This critique is thus different from and, it could be argued, more extensive than that of Marxism. It goes 'far beyond the usual Red interpretation and delivers a civilizational criticism rather than a simple criticism of a particular mode of production' (Friberg and Hettne 1985, p. 218). It demands that non-material needs be addressed, and judges 'the so-called developed societies' wanting: they 'are to a certain extent responsive to our material needs, but when we introduce the concept of non-material needs, the whole picture is turned upside-down' (p. 219).

In place of evolutionary developmentalism, Friberg and Hettne offer a normative concept setting out the desired direction of development which is radical and humanistic, based on the concept of 'self-realization', the 'full realization of the individual in every aspect of its being' (Friberg and Hettne 1985, p. 216). This definition rules out a number of other goals, such as 'economic growth, technological development, scientific growth, social revolution, serving a particular class, serving the national interest, the state or God' (p. 216). It allows an optimistic view of human nature, despite the record of destructiveness, because this is second nature. It is society that is deviant, not individuals: 'when there are no outlets for love and creativity, man becomes a stunted creature characterized by a life-inhibiting syndrome of sado-masochism, greed, egocentricity, and callousness' (p. 219).

Beyond the primary definition of development in terms of fundamental human needs, Friberg and Hettne outline 'Green' principles of 'endogenous development' (Friberg and Hettne 1985, p. 220). These are, first, that the social unit of development should be a culturally defined community, and its development should be rooted in its values and institutions; second, self reliance, so that each community 'relies primarily on its own strength and resources'; third, social justice; and fourth 'ecological balance', implying an awareness of local ecosystem potential and local and global limits. Obviously this formulation reflects a great deal of the thinking already familiar from the World Conservation Strategy and the Brundtland Report. For example, it picks up again the ecologically-based concerns of environmentalism, and indeed Friberg and Hettne acknowledge ecology as 'one main source of inspiration behind the Green view' (p. 224). Although familiar, the distinctiveness of this green view is stressed, together with the 'paradigmatic blindness' of both 'red' and 'blue' views of the world with regard to ecology (p. 225).

This idea of 'endogenous development' is, they suggest, 'completely at odds with dominant universalist thinking' (Friberg and Hettne 1985, p. 220). Development is to be sought in a country's own ecology and culture, not in the supposed 'model' of a developed country. Furthermore, as development is to be through 'voluntary cooperation and autonomous choices by ordinary men and women' (p. 221), the unit of development cannot be considered to be the state, but people and groups of people defined by culture, indeed 'natural communities' (p. 221). Similarly, their green view of self-reliance is neither the established idea of national self-reliance (for example included in the Cocoyoc declaration), nor autarchy but self-reliance at different scales within society. Similarly, it is argued that the green notion of social justice goes beyond the established idea of redistribution with growth, and monetary compensation for marginalization and alienation, to embrace access to wealth, knowledge, decision-making and meaningful work.

Having abandoned the notion that modernization is the goal at which humanity aims, Friberg and Hettne pick up the ideas of the world-system from Wallerstein, and speak of 'the modern project'. This describes the modern world-system, dominated by capitalism, the state and the 'scientific-educational system' (Friberg and Hettne 1985, p. 231). It was embarked upon first in Western Europe, and was imposed upon the periphery through geographical expansion and socio-economic penetration. That penetration implies the decline and disintegration of 'natural communities', and theoretically the modern project will eventually eradicate all pre-capitalist social formations. Friberg and Hettne distinguish between two strategies, the 'red road of continued modernization', which involves allowing the world-economy to expand to a ceiling, and then seeing it replaced by a socialist world-government, and a green strategy of 'demodernization'. This would involve gradual withdrawal from the modern capitalist world-economy and the launch of a 'new, non-modern, non-capitalist development project' (p. 235). This new project would be based on 'the "progressive" elements of pre-capitalist social orders and later innovations' (p. 235), the exploitative and dehumanizing element in some small-scale pre-capitalist orders being recognized.

This model therefore involves a dualism between modern and non-modern institutions, a modern project ('red' and 'blue') which aims at control, expansion, growth and efficiency and is legitimized by evolutionist thinking, and set against this (and incompatible with it) a green project aiming at human development. The green project draws its strength from an alliance of three groups. The first consists of 'traditionalists' who wish to resist capitalist penetration in the form of state-building, commercialization and industrialization. These will be mostly in the global periphery and will include 'non-Western civilisations and religions, old nations and tribes, local communities, kinship groups, peasants and independent producers, informal economies, feminist culture etc.' (Friberg and Hettne 1985, p. 235). The second group is 'marginalised people', including the unemployed, the mentally ill, handicapped people and people in de-humanizing jobs who have lost 'a meaningful function in the mega-machine' (p. 264) through pressures for increased productivity, rationalization and automation. The third group consists of the 'post-materialists' who dominate western environmentalism (Cotgrove 1982), 'young, well-educated and committed to post-materialist values' (Friberg and Hettne 1985, p. 236).

This attempt to define a common constituency for green action is in some ways the most interesting aspect of the green project outlined by Friberg and Hettne. The key element is the circumvention (or in some cases subversion) of the nation-state (which they see as 'one of the greatest obstacles to "the Greening of the world"', Friberg and Hettne 1985, p. 237) by counterpoint movements. Although Friberg and Hettne discuss constructive projects carried out by spontaneous networks and voluntary organizations, particularly in the context of India, in the Third World counterpoint movements are primarily movements of protest.

The modern project is maintained by 'modern elites' who are in the ambiguous position of increasing involvement in the world-economy to increase their resource base, and at the same time foster the development of the political, military and bureaucracy of the nation-state as a power base. Against this, they discuss the extent and potential of non-elite counter mobilization, or 'counterpoint movements', which they call 'non-party politics'. This they see as 'the only force that can counter the power of the modern elite' (Friberg and Hettne 1985, p. 244). Reactions to capitalist penetration can be violent and revolutionary, but non-party politics 'expressing the citizen's concern with the exploitation and corruption that typically form part of the functional system' (p. 244) is rarer in the Third World. It involves 'spontaneous organisation and networks emerging from the economic and political crisis' (p. 242), and results from both the inherent strength of those groups and networks and a weakening of the power of the state.

The corollary of this Third World picture is the application of the green project in the industrialized West. Again, this involves a modern/non-modern dualism, and the formation (primarily so far by a post-materialist elite) of a 'new counter-structure to the formal-rational structure of modern society' (p. 250). Friberg and Hettne draw in 'a whole new set of social movements trying to articulate a third (green) vision of the world' (p. 248). These include 'the ecological, solidarity, peace, feminist, communal, regional, youth, personal growth and new age movements' (p. 248), which they see as 'part of a worldwide current defending natural communities and searching for alternatives to modernization' (p. 248). Many different 'alternative' movements from the 1950s anti-bomb protests onwards 'all converge towards the small-scale logic of functionally integrated communal societies based on direct participation and self-management' (p. 258), in which capitalist growth, central bureaucracy and 'techno-science' has a small place. This is the essential non-modern 'green' vision.

Interestingly, the labour movement is more or less dismissed as an element in this counter-structure, arguing that it has become 'wedded to the large-scale industrial system' (Friberg and Hettne 1985, p. 253), and hence reproduces the features of the modern system: 'The old labour movement has chosen the modern road and it has now become incorporated in the formal-rational complex, and is now losing its momentum as a creative force' (p. 259). Automation and deindustrialization will, however, 'shake the working class to its very foundations' (p. 254), leaving part of it open to the demodernization project of the green movement. This would seek autonomy by mass withdrawal from the formal system - through informal organizations and decentralized networks, personal development and direct participation. Obviously green thinking about development of the kind discussed by Friberg and Hettne represents a considerable advance in sophistication on that dominant in writing on sustainable development. Many of the established ideas from less theoretically developed formulations such as the World Conservation Strategy are present in their work, some in an expanded form (for example the notion of person-controlled change) and others vestigial (for example the whole area of genetic conservation and wildlife protection). This green thinking is innovative in that it does address the major issues of radical development theory, and in doing so it at least opens up the possibility of dialogue.

Table 4.1 Comparison of red and green strategies of development

	Red strategy	*Green strategy*
Oppressive system	The capitalist economy	The technological culture
Enemy	Capitalists	Technocracy
Social vision	Socialist society	Communal society
Method	Macro-revolution, socialism from above by working class revolution	Micro-revolution, small groups withdraw from system to defend autonomous ways of life
Spatial frame	No socialist islands in a capitalist sea	Local experiments, liberated zones
At stake	Material interests, collective identity, ownership	Existential needs, personal identity, autonomy
The new person	Social transformation before personal	Simultaneous personal and social transformation
Leadership	Intelligentsia	Postmaterialistic elite
Social base	Working Class	Marginalized people
Institutional base	Big industrial sites	Small neighbourhood communities
Organization	Centralized, formal	Decentralized, informal
Ideology	Abstract, rational,	Concrete, intuitive, open

Source: Friberg and Hettne (1985)

Friberg and Hettne try to demonstrate the distinctiveness of 'green' and 'red' development strategies (see Table 4.4). These distinctions are not entirely successful. What validity they have depends on the simplistic and one-sided caricature of 'red' or Marxist thinking which they offer. Having pointed out the existence of 'green' strands within radical thought, for example utopian socialism, they proceed to ignore these in claiming the distinctiveness of 'green' ideas. There are, however, indeed important strands of radical thought that offer a significant 'alternative' critique of development, moreover one that is at least as 'green' as it is conventionally 'red'.

Radical critiques of environmentalism

In understanding how radical 'green' ideas relate to Marxist thinking it is first necessary to examine radical critiques of the environmentalism from which it springs (e.g. Enzensberger 1974, Sandbach 1978). In this critique it is probably the 'futures' debate which has attracted most attention, both in the directly 'neo-Malthusian' arguments about overpopulation in books such as *Population, Resources and Environment* (Ehrlich and Ehrlich 1970) and *The Population Bomb* (Ehrlich 1972), and the global computer models of Forrester (1971) and Meadows *et al.* (1972). Criticism came not only from economists such as Beckerman (1974) and Simon (1981), and scientists (whom Cotgrove (1982) might describe as 'cornucopian') with particular faith in the benign possibilities of technological change (e.g. Maddox 1972), but also from radical thinkers. Thus Harvey (1974) argues bluntly that 'the projection of a neo-Malthusian view into the politics of the time appears to invite repression at home and neo-colonialist policies abroad. The neo-Malthusian view often functions to legitimate such policies and, thereby, to preserve the position of a ruling elite' (p. 276). The 'new barbarism' (the phrase is from Commoner 1972) of neo-Malthusianism, probably most firmly identified with Hardin's 'lifeboat ethics' (Hardin 1974), has been called 'ecofascist' (Enzensberger 1974, Bookchin 1979, Pepper 1984).

Certainly, neo-Malthusianism has a heavy political content (Enzensberger 1974), emerging as colonial rule came to an end in the Third World. It was taken up particularly in the USA, where ideas moved rapidly from the theoretical to the political arena. Such ideas were inevitably ideological. Their claimed social neutrality, stemming from their roots in natural science, was simply 'a fiction' (Enzensberger 1974, p. 9). Indeed, Enzensberger saw ecology overwhelmed by its inability to theorize sensibly about society, and taking shelter in global projection, 'a surrender in the face of the size and complexity of the problems which it has thrown up' (p. 17). Quite simply,

> In the case of man, the mediation between the whole and the part, between subsystem and global system, cannot be explained by the tools of biology. This mediation is social, and its explication requires an elaborated social theory and at least some very basic assumptions about the historical process.
>
> (Enzensberger 1974, p. 17)

Politically, the impact of neo-Malthusian ideas is clearly visible in the USA. The 'positive programme' introduced rather apologetically by Ehrlich and Ehrlich (1970) at the end of their book is an American vision, based on a view of the world where American political and economic hegemony is assumed. It discusses what the American government could and should do nationally and internationally to control population. Other organizations, for example The Environmental Fund (of which Paul Ehrlich and Garrett Hardin were founder members) and various groups dedicated to population control such as ZPG and Planned Parenthood, focused on the population content of USAID giving, with significant impacts on USAID policy. These have only been overturned by attacks in the 1980s by the anti-abortion lobby in the USA, ironically of course from a similar position to the right of the political spectrum as the original proposers of population control.

Since the early 1970s considerable effort has been put into modelling 'futures' after the style of *The Limits to Growth* (Cole 1978), yet Sandbach (1978) argues that enthusiasm for the more pessimistic aspects of the debate rapidly died down. One reason for this was the selective and partial take-up and promotion of ideas. Enzensberger predicts attempts by capital to control the environmental movement through an 'eco-industrial complex', pointing out how all environmentalist ideas have to undergo 'processing through the sewage system of industrialised publicity' (Enzensberger 1974, p. 7). Golub and Townsend (1977) develop the same kind of argument, describing the way in which the Club of Rome picked up the calls for global stability from *The Limits to Growth* but not the notion of zero growth.

Certainly of all the ideas current in the early 1970s, only that of population control caught on in terms of policy, again, Sandbach suggests, because it 'did not threaten the fabric of advanced capitalist countries. On the contrary, population control, especially in the less developed countries, would ensure a greater slice of the resources cake in the future' (Sandbach 1978, p. 29). Concern for population growth in the Third World and its supposed 'carrying capacity' has persisted

(Kirchner *et al.*1985). Westing (1981), for example, argues that the maximum sustainable world population is 2 billion, while a major FAO study using computer models to estimate population supporting capacities suggests severe problems in a number of countries (Higgins *et al.* 1982)

The neo-Malthusian environmentalist vision of the Third World's future is grim: many of the underdeveloped countries would 'never, under any conceivable circumstances, be "developed" in the sense in which the United States is today. They could quite accurately be described as the never-to-be-developed countries' (Ehrlich and Ehrlich 1970, p. 2). This pessimistic thinking has been picked up and explicitly applied to 'ecodevelopment', for example by Riddell (1981). Thus May speaks of the 'dead end societies' of the Third World: 'there is no prospect of change in the Third World that would substantially improve the lives of more than a few people' (May 1981, p. 226). Interestingly, these ideas are not confined to the 'catastrophists' among environmentalists. Maddox takes a similarly line while attacking the 'eco-doomsters'. He argues against the 'bland assumption that developing countries may somehow be unable to follow the path of development along which advanced countries have travelled' (Maddox 1972, p. 4). The Marxist case against this naïve environmentalism, along with its hidden ideology and political implications, is debated in some detail by Pepper (1984). Environmentalism takes an idealist rather than materialist stance, it is ahistorical and fails to recognize that resources and indeed ecological problems are socially defined and therefore conditioned by mode of production. Because of this, environmentalism offers an essentially determinist analysis based on the principle of unchanging limits on human action, and a pessimistic view of the potential impact of social reform. This generates, he suggests, the deeply conservative ideology and reactionary and repressive politics of 'ecofascism'.

Reds and greens in development

Obviously not all 'green' ideas can be dismissed as easily as this. As Pepper points out, there is 'a considerable degree of coincidence' between some environmentalist visions of the future and those of social anarchism or anarcho-communism, and utopian socialism (Pepper 1984, p. 196). He goes on to define the elements of 'ecosocialism' involving the redefinition of needs, the redistribution of resources, the reassessment of the industrial mode of production attention to the alienating effects of work structures. The 'left-wing radicalism' of ecosocialism would go further, and focus on new forms of production which eschew private ownership in favour of social justice, and new forms of social order which 'eliminate alienation, state control, and centralisation' (p. 197).

Obviously, therefore, to an extent radical green thinking does make sense in terms of anarchist or Marxist thought. But how distinctive and different is it? There are two elements to be considered here, first the extent to which radical thinking has embraced environmentalist thinking, and second how far it comprehends it and makes claims of an independent 'green' paradigm meaningless. Certainly the 'environmental crisis' is acknowledged to have been an underplayed element in socialist thinking. Given the 'confused alliance' between environmentalism and diverse political movements, Enzensberger considers it unsurprising that the left in Europe has simply 'incorporated certain topics from the environmental debate in the repertory of anti-capitalist agitation', while remaining sceptical and aloof from environmentalist groups (Enzensberger 1974, p. 9). At that stage its critique was primarily ideological, but Enzensberger demonstrated the reality of scientific problems lying behind the bourgeois packaging of the environmental movement, and reiterated the 'commonplace of Marxism' which environmentalists highlighted, that the capitalist mode of production 'has catastrophic consequences' (p. 10). The editorial in that 1974 volume of *New Left Review* asserts that 'to identify and combat these has become a central scientific and political task of the socialist movement everywhere'.

The most frequently cited work which links 'green' and 'red' thinking is probably Bahro's *Socialism and Survival* (1982). Bahro occupies a very distinctive position both personally and intellectually, having left East Germany after the publication of *The Alternative in Eastern Europe* (Bahro 1978) and having become involved in the Green Party in West Germany. Of that involvement he said 'I believe that the ecological crisis will bring about the end of capitalism, and that is why I have associated myself with the Greens in West Germany' (Bahro 1984, p. 119). Bahro is most easily (but not entirely satisfactorily) classified as a Marxist, although his ideas stray far beyond the folds of Marxism, let alone Marxist orthodoxy.

In *From Red to Green*, Bahro suggests that it could be said that he had 'left Marx behind', quarrying from Marxism those elements he found useful and moving from scientific to utopian socialism and from 'a class-dimensional to a populist orientation' (Bahro 1984, p. 220). In doing so Bahro of course moves closer to Bookchin's anarchist thinking, and becomes distinctively 'green'. He suggests 'we have now reached the point where humanity has to find a new stable life-form in which its forward development is an inward journey rather than an external expansion' (p. 221). His ideas also become heavily influenced by Western environmentalist thinking. In *From Red to Green* he stresses the fundamental nature of ecological limits. The ecological crisis brings 'questions onto the agenda which were already there before the first class society took shape' (p. 148), and he suggests 'the time has come when the utopian communist and socialist visions are no longer utopian. We have reached the limit. Nature will not accept any more, and it's striking back' (p. 184).

The ideas in *Socialism and Survival* (Bahro 1982) are reviewed by Redclift, who suggests that environmental consciousness 'offers a more radical and enduring critique of underdevelopment than any other currently on offer' (Redclift 1984, p. 55). He discusses the challenge of Bahro's radical green thinking to both Marxism and environmentalism. Bahro's 'red-green' ideas in fact offer a double critique, of capitalism and what he terms 'actually existing socialism'. In particular, he attacks industrialization, both within the capitalist and socialist systems. He argues that 'the old industrial system is at an end: the increase in material consumption and production, with the inbuilt waste, pollution and depletion of resources, has reached its non plus ultra and is enough to destroy us in a few generations' (Bahro 1984, p. 179). He also claims that there is a need to dismantle 'the huge structures we have built' and to raise the question not simply of the need to reform the kind of products made, but the 'reproductive process itself' (p. 147). He visualizes a phase of contraction and of dissolution of the world market, replacing its material infrastructure with an 'informational structure' (p. 180).

Established development strategies for the Third World through increasing trade and industrialization Bahro sees as 'a tunnel without exit' (Bahro 1984, p. 211). He believes that 'continued industrialism in the Third World will mean poverty for whole generations and hunger for millions. The prospect of proletarianising humanity is a horrific vision' (p. 184). Poverty in eighteenth- and nineteenth-century Europe was 'made bearable' by the prospect of escaping it through exploitation of the periphery. However, 'for the present periphery there is no further periphery to be exploited, no way of attaining the good life of London, Paris or Washington' (p. 211). Bahro's views on industrialization are to an extent shared by Sutcliffe. In 1972, Sutcliffe argued for industrialization in large scale units under autarchic conditions. In 1984 he reassessed this position, arguing against the assertion that capitalism could 'develop' the whole world to the level of the industrialized West, thus preparing the way for an advanced socialist society (Sutcliffe 1984). In this Sutcliffe reflects the sharp debate within radical thought about the place of industrialization, and the supposedly 'progressive' role of imperialism and Third World industrialization. Lenin (1972) saw imperialism as the final stage of capitalism, both 'parasitic and decadent' (Corbridge 1986). Warren challenged this, and argued that rapid industrialization was occurring in the Third World. The fact of this stands in contrast to 'dependency' thinking (which argues that very little is happening), and against populist arguments that what is happening is far from desirable (e.g. Lipton 1977).

Sutcliffe's point in his 1984 paper is that the economic structure of advanced capitalism is simply 'inappropriate as a material basis for a socialist society' (Sutcliffe 1984, p. 132). Capitalism produces inappropriate products (cars, weapons, obsolescent goods), and it uses the wrong methods, methods that are centralized, de-skilled, totalitarian and alienating. In doing so it has created its image in a 'centralised, statist and bureaucratic' view of socialism. By contrast, Sutcliffe points to certain strengths in the critiques of industrialism by populists and intermediate technologists. He urges a recapture of utopian traditions by socialist thinking on industrialization and development, a search 'for a more humane alternative to economic development than the rocky path represented by actually existing industrialisation' (Sutcliffe 1984, p. 133).

Other approaches from within a radical tradition address issues which also engage with the 'green' critique of development. One example is Caldwell's discussion of 'homeostasis' (described as 'sustainable eco-economic equilibrium') as an alternative to both underdevelopment and 'overdevelopment' (Caldwell 1977, p. 139). He puts forward the experiences of China, North Korea, Vietnam, Laos and Cambodia as examples from which developing countries (particularly in Asia) might learn, and then discusses the possibilities of 'transcending overdevelopment' in Great Britain. This is an interesting treatment, partly because there is so little literature available on political economy and environment within the socialist world. What work exists highlights the existence of environmental problems in the USSR (e.g. Mote and Zumbrunne 1977, Hollis 1978, Pryde 1972, Komarov 1978) or China (e.g. Smil 1984) under conditions of what Bahro (1984) would call 'actually existing socialism'.

Redclift suggests that 'the environmentalist consciousness developing in Europe, shorn of the parochialism evident in the American and British literatures, offers a more radical and enduring critique of underdevelopment than any other currently on offer' (Redclift 1984, p. 55). He calls for 'a fundamental revision of Marxist political economy, to reflect the urgency of the South's environmental crisis' (p. 18), and in his subsequent book (Redclift 1987) tries to do something to flesh out this requirement. Redclift is by no means alone in drawing attention to the relative neglect of environmental issues in radical social studies, the same point being made (and tackled) very effectively for example by Blaikie (1985) and Blaikie and Brookfield (1987). In passing it is worth noting that the accusation is no longer entirely fair, the theoretical framework for a Marxist theory of nature being discussed *inter alia* by Burgess (1978), and Smith (1984), and an anthology of writings by Marx and Engels on ecology being provided by Parsons (1979).

Nonetheless, it is clear that there is now engagement between radical development theory and environmentalism, even if it is still rather limited. Does this imply that all radical green ideas can be subsumed under this 'ecosocialist' umbrella? This case is certainly made. Pepper (1987) replies to Owens' (1986) discussion of 'green politics' in the British context, and suggests that the real question is not whether the established parties are approaching a 'new paradigm of the Green movement', but whether environmentalists are abandoning as facile their claim to be above politics, and are realizing that the 'old politics' of left and right are instrumentally relevant in achieving their desired new society. To him, radical greens are 'red greens'.

The same sense of frustration at the claims of greens to be a truly radical alternative is shared by Amin. He challenges the claim that Greens lie 'beyond the debate between traditional right and traditional left' (Amin 1985 p. 272). He argues that 'the Greens - mistakenly - do not acknowledge that their counter-society is in fact nothing more than the communism espoused by Marx' (p. 273). He maintains that green ideas are not new, but have roots (as we have seen) in peasant communism (itself 'not specific to any particular society') and utopian socialism: 'the revolt of the oppressed and exploited has always produced the ideology of a just, equal and mutual counter-system' (p. 273). In particular, Amin suggests that 'many of the themes which sustain the "green" movement have their origins in Maoism' (p. 178), and he argues that the Greens pass over this heritage, and the Chinese experience (ideas such as self-reliance, decentralization, and delinking), too lightly.

The Greens do not constitute 'an organised and homogeneous movement' (Amin 1985, p. 279) and Amin argues that it is unwarranted to group together the huge diversity of 'alternative' movements under one banner. Thus the feminist movement 'cannot be annexed to the logic of the Greens any more than to any other' since some 'are perfectly compatible with capitalist logic, while others subscribe to the logic of socialism' (p. 279). Amin sees the rise of green thinking as a reflection of the failure of socialism:

> The appearance of political currents bearing the name green is no surprise. It is the manifestation of the failure of the strategies operated by the European left over several decades. As such it demands a serious self-criticism, and represents a positive phenomenon.
>
> (Amin 1985, p. 280).

Green thinking itself he dismisses as 'a form of religious fundamentalism' (p. 281) akin to that of Islam, appearing in the Third World because of wrong-headed communist strategies. Although theoretically far removed from Pepper's concern with the environmental policies of the British left, for Amin too truly radical green thinking must be 'red-green'.

Radicalism, development and environment

It is clear that 'green' thinking about development, like its unradical alter ego in sustainable development, is heterodox and eclectic. However it is clear that there are radical critiques of the development process which are beginning to comprehend political economy. These have the makings of a coherent understanding of how society and nature relate which goes far beyond the simple oppostionism of environmentalism. They also have the potential to create agendas for action which transcend the limited reformism of sustainable development thinking. In a sense, that thinking asks some enormously important questions, but provides no answers beyond reform of the procedures and organization of development planning. Their radicalism is restricted, untheorized and naïve.

In order to move beyond this, it is necessary to move outside environmental disciplines, and outside environmentalism, to approach the problem from political economy and not environmental science. There is a great power in an approach to the understanding of environmental aspects of development which uses both natural science and social insights, just as Blaikie (1985) attempts to integrate natural and social sciences in the analysis of environmental 'problems' such as soil erosion.

Even this does not go far enough, because it fails to challenge the ideology of development itself. All relations between environment and people are political, just as all development is ideological. In a preface to the 1979 reprint of his paper 'Ecology and revolutionary thought', (originally written in 1965), Murray Bookchin argues that 'the domination of human by human lies at the root of our contemporary Promethean notion of the domination of nature by man' (p. 21). The way people relate to their environment - as well as the way they understand it - is created by culture, and bounded by social relations, by structures of power and domination. Development itself is a product of power relations, of the power of states, using capital, technology and knowledge, to alter the culture and society of particular groups of people. States co-opt cultures while the world system engages indigenous economies. We call the result of this process development.

Development is therefore about control, of nature and of people. It is at this level that sustainable development thinking is seen to be so deeply unradical. For although it is a major strength of a great many of the environmental critiques of development (including many of those that are naïve and limited in scope) that they adopt an essentially moral critique of the development process, they offer in its stead simply another version of the same paradigm. We have seen how sustainable development depends on arguments about capitalist mutualism. It is one which is deeply dependent on a vision of managerialism drawn from ecology and, in turn, from systems theory. McIntosh (1985) gives an account of the growth of systems theory in ecological science. Its methodology, drawn from the sciences of engineering and thermodynamics, offers an ideology of control which has attained a wide currency within environmentalism. It has an important position in the thinking of environmentalists about development. The fact that the ideological component of systems theory is often cryptic does nothing to lessen its significance (Gregory 1980). Similarly, the view of development as essentially a managerial process, where reform of procedures will ensure some 'optimal' outcome, fails to address the central issue of the ideology of the 'modern project' (to use Friberg and Hettne's 1985 phrase).

Development creates losers as well as winners. It is something that is done to people, usually by governments, often backed by ideologies and resources from elsewhere. To the recipient, the target, this 'development' holds many terrors, many costs as well as benefits. States make decisions on behalf of people regarding proper balances of costs and benefits, and about where those costs should be borne and those benefits enjoyed. Sometimes (as environmentalists devastatingly point out) the two do not balance. Even more often the costs and benefits are unequally shared.

In a sense then development is a double-edged phenomenon, a tiger which governments ride, and to which they attach (sometimes by force, Crummey 1986) the lives of individuals and groups of people. This is a perspective which surfaces in some environmentalist thinking on development (such as voiced concern for indigenous people, e.g. McNeely and Pitt 1987), but is rarely included in theoretical work. It is reflected most closely in anarchist thinking, for example in the work of people such as Kropotkin and his ideas of social anarchism or anarcho-communism (Kropotkin 1972, 1974, Breitbart 1981, Galois 1976), or in the work of writers such as Bookchin (1979), and Roszak (1979).

The central point here is the importance of viewing development from the perspective of the individual. Kropotkin believed 'that true individualism can only be cultivated by the conscious and reflective interaction of people with a social environment which supports their personal freedom and growth' (Breitbart 1981, p. 136). Pepper (1984) points out the difference between this starting point in the individual and that of environmentalist thinking such as *Blueprint for Survival* (Goldsmith *et al.* 1972), which suggests an ecological imperative for action to achieve utopia, the human dimensions of which are secondary. Bookchin argues that 'in the final analysis, it is impossible to achieve a harmonisation of people and nature without creating a human community that lives in a lasting balance with its natural environment' (Bookchin 1979, p. 23).

Bookchin bases his arguments quite extensively on ecological ideas, suggesting for example that the 'critical edge' of ecology is due not to the power of human reason, but 'to a still higher power, the sovereignty of nature' (Bookchin 1979, p. 24). He sees ecology as one science which might avoid assimilation by the established social order, and he mourns the absorption of environmentalists into governmental institutions, and warns that 'ecological dislocation proceeds unabated, and cannot be resolved within the existing social framework' (p. 22). Ecology is integrative and reconstructive, and this 'leads directly into anarchist areas of social thought' (p. 23). Thus:

> The anarchist concepts of a balanced community, a face-to-face democracy, a humanistic technology and a decentralised society - these rich libertarian concepts - are not only desirable, they are also necessary. They belong not only to the great visions of a human future, they now constitute the preconditions for human survival.
>
> (Bookchin 1979, p. 27)

To Bookchin the 'ecological crisis' was just part of a larger malaise: 'Humanity has produced imbalances not only in nature but, more fundamentally, in relations amongst people and in the very structure of society'. Therefore, 'what we are seeing today is a crisis in social ecology', in which Western society is 'being organised round immense urban belts, a highly industrialised agriculture and, capping both, a swollen bureaucratised, anonymous state apparatus' (Bookchin 1979, p. 25). Bookchin's 'ecological anarchism' is therefore essentially anti-industrial, anti-bureaucracy and anti-state. In some ways it covers similar ground to Schumacher's *Small is Beautiful* (1973), but for Bookchin that approach demands adaptation to the norms of society, not the 'revolutionary opposition' he believes is required.

Some writers who are broadly in the anarchist tradition lie further back on this axis between conformism and revolution. Thus Roszak, in *Person/Planet*, points out that 'we need more to guide us than a slogan like "small is beautiful"' (Roszak 1979, p. 306). Roszak argues that as long as western society remain locked into 'the orthodox urban-industrial vision of human purpose', there is no hope that poverty, and the injustice it brings with it, can be 'more than temporarily and partially mitigated for a fortunate nation here, a privileged class there (Roszak 1979, p. 317). This view obviously falls short of the revolutionary position advocated by Bookchin, but it contemplates more specific and perhaps more radical restructuring of the organization of society and of production.

In development the domination of nature is part of wider political and economic processes. It is impossible to understand the relation between development and nature without considering and comprehending political economy. However, once we escape from the straightjacket of evolutionary thinking about 'development', which argues that it involves a progression towards 'better' states, it is possible to start to consider anew the persistent questions raised by environmentalist critiques of development practice. Specifically, these are how poverty and environmental quality relate to each other, how development actions impact the environment and why the reform of development practice proves to be so difficult to achieve.

These questions are addressed in the rest of this book. It discusses the implications of the political and economic context of environmental degradation (Chapter 5), looking specifically at the issue of desertification. It examines the question of pollution associated with tropical development, and the nature of the environmental impacts of development in rainforest environments, and in tropical water resource developments (Chapter 6). It considers both the failings of existing reformist attempts to place environmental issues high on development agendas, and the practical constraints on their successful implementation (Chapter 7). It then returns to the central question of the impact of development on people, and argues that people and poverty must be at the centre of thinking about environment and development. Above all, it tests the conflict between radicalism and reformism, exploring the relations between people and environment in development and assessing the limitations of the reformist model of sustainable development.

5

Development and Environmental Degradation

Poverty, environment and degradation

Poverty and environment are linked in a close and complex way. Poor people live in and suffer from degraded environments, and very often they create environmental degradation because their poverty forces them to do so. It is obvious, but worth repeating, that farmers in the highlands of Ethiopia or Nepal do not farm steep and eroded hillsides through perversity but through necessity (Blaikie 1988). On the other hand, degraded environments themselves create poverty. Thus drought in the Sahel or pollution from mine tailings in tropical rivers exacerbate the exposure of the poor. Very often those affected have neither the freedom to stop causing degradation, nor the opportunity to move elsewhere. Poverty and environmental degradation form a trap from which there is little chance of escape. This linkage between poverty and environmental degradation provides perhaps the clearest demonstration of the centrality of social, political and economic issues in questions of environment and development. It is on the environments in which the poor live and from which they draw their sustenance that concern about sustainable development has to focus.

The reciprocal links between poverty and environmental degradation forces what Blaikie describes as the 'desperate ecocide' of the poor, small producers who 'cause soil erosion because they are poor and desperate, and for whom soil erosion in its turn exacerbates their condition' (Blaikie 1985, p. 138). Obviously the environment is not neutral in its effects on the poor; environmental quality is mediated by society, and society is not undifferentiated. Access to and the distribution of environmental 'goods' (be they cultivable land, fuelwood or clean air) is uneven. Blaikie argues for a political-economic understanding of the phenomena of environmental degradation:

> It is when the physical phenomenon of soil erosion affects people so that they have to respond and adapt their mode of life that it becomes a social phenomenon. When this response affects others and brings about a clash of interests ... it becomes a political problem as well.
>
> (Blaikie 1985, p. 89)

Blaikie's central argument is that it is vital to see the links between environment, economy and society. However, each of these things is itself highly complex. It is 'development' which very often provides that dynamic. In a crude sense, development both affects and is affected by environmental quality. Thus Blaikie and Brookfield argue 'land degradation can undermine and frustrate economic development, while low levels of economic development can in turn have a strong causal impact on the incidence of land degradation' (Blaikie and Brookfield 1987, p. 13). More importantly, because the 'development process' (Mabogunje 1980) involves the transformation of social and economic relations, it relates to the ways in which individuals and groups within a society experience their environment, and the ways in which they use it.

The links between environmental degradation, poverty and development are seen most clearly in the case of arid land desertification. This has become one of the recurrent themes of environmentalist writing on the Third World (e.g. Grainger 1982, Brown and Wolf 1984, Timberlake 1985). Desertification shares with one or two other issues (notably the destruction of tropical rainforest and related issues of the greenhouse effect and global warming) the distinction of being widely assumed to prove the unsustainable nature of human relations with the environment on a global scale. These are what are described in *Our Common Future* as 'environmental trends that threaten to radically alter the planet, that threaten the lives of many species upon it, including the human species' (Brundtland 1987, p. 2). Just as the 1972-4 Sahelian drought led to an international awareness of desertification, and hence to the United Nations Conference on Desertification (UNCOD) in Nairobi in 1977 (see Chapter 3), the continuing work of UNEP and the persistence of Sahelian and African drought in the 1980s has kept attention focused on desertification issues (see for example Karrar 1984, Mabbutt 1984, Tolba 1986, Rapp 1987).

Three aspects of this interest in desertification are of particular importance. First is the way in which there can be such agreement about the nature, location and extent of the problem while there is such an unclear definition of exactly what desertification is. Concern about arid land degradation is long established, particularly in Africa, and the perception of such degradation is unmistakeably the fruit of particular ideologies. Before ecological or other knowledge is brought to bear, ideology has a major influence on what people see.

Secondly, the issue of desertification is an issue perched squarely on the division of the natural and social sciences. Some seek to explain its cause in terms of environmental change (usually climatic change), while others stress the influence of society and governments, of political economy and colonialism. In this way desertification provides an important illustration of wider theoretical debates about environmental aspects of development.

Thirdly, desertification provides an example of the ways in which developers respond to perceived environmental problems. It therefore makes it possible to examine the relationship between development and environmental degradation, to consider the role of development as both a response to perceived environmental problems and yet also a creator of them.

Certainly there is little doubt of the importance placed on desertification in literature on sustainable development. Brundtland states that 'each year another 6 million hectares of productive dryland turns into worthless desert' (Brundtland 1987, p. 2). The *World Conservation Strategy* sees desertification as 'a response to the inherent vulnerability of the land and the pressure of human activities' (IUCN 1980, para 16.9). It describes the way in which 'overstocking' has severely degraded grazing lands in Africa's Sahelian and Sudanian zones', with farmers 'moving onto land marginal for agriculture' displace pastoralists onto land marginal for livestock rearing, while in the Himalayas and Andes 'too many improperly tended animals remove both trees and grass cover ... and erosion accelerates' (para 4.13).

This concern is mirrored by the abundance of stark statistics on the extent and severity of desertification, particularly from UNEP. Thus in 1983, UNEP estimated that about 35 per cent of the terrestrial globe is vulnerable to erosion (some 4500 million hectares), and this supports about a fifth of the world's population, some 850 million people. Of this area, about 80 per cent of rangelands (3100 million hectares) are actually affected by 'overuse', plus 335 million hectares of rainfed croplands (Tolba 1986). Of the 4.5 billion hectares at risk of desertification globally, 75 per cent is already moderately desertified and 30 per cent severely or very severely desertified. The extent and severity of desertification was increasing everywhere in the Third World (Mabbutt 1984). Although these dire statistics are often repeated, the factual base from which they are drawn is still remarkably slim. There are relatively detailed studies for certain areas, for example the Nile Basin, for which studies have attempted to estimate present rates of water and wind erosion and land salinization (Kishk 1986). However such studies are few and far between. Despite the advent of monitoring methods such as satellite remote sensing (see for example Tucker *et al.* 1985, 1986, Justice *et al.* 1985, Prince 1986), data are still lacking on the extent of land degradation on a global scale.

Ormerod (1978) points out the need for quantitative techniques for assessing large-scale loss of fertility if clear-cut policy decisions are to be made in potentially desertified regions. Attempts to assess the extent of desertification use indicators such as the growth and encroachment of moving sand, the deterioration of rangelands, the degradation of rainfed croplands, waterlogging and salinization of irrigated areas, deforestation (usually specifically in arid environments) and declining ground or surface water supplies (Mabbutt 1984). Ormerod (1978) outlines possible approaches based on studies of soil loss (or erosion), for which the US Bureau of Reclamation's Universal Soil Loss Equation provides the most widely recognized basis, and on studies of loss of the organic content of soils. Verstraete (1986) suggests albedo, vegetation cover, soil depth, livestock composition, human health and average revenues might all be taken as indicators of desertification. Notwithstanding the relative ease with which the requirements of desertification monitoring can be set out, quantitative field studies are unusual. Perceptions of the severity of desertification are to a large extent self-confirming, but have a scanty scientific base.

Furthermore, despite the fact that the word desertification is now widely known and freely used by environmentalists, politicians and policy-makers alike, there is a surprising confusion over what exactly is meant by desertification (Warren and Agnew 1988). Desertification embraces the phenomena of drought, soil erosion, the destruction of vegetation cover, and the degradation of human living conditions, but it cannot be reduced to these things alone (Verstraete 1986). Grove defines it as 'the spread of desert conditions for whatever reasons, desert being land with sparse vegetation and very low productivity associated with aridity, the degradation being persistent or in extreme cases irreversible' (Grove 1977, p. 299). The UNCOD Plan of Action on desertification defines it as 'the diminution or destruction of the land', which 'can lead ultimately to desert-like conditions' (Mabbutt 1984, p. 103), while Warren and Maizels suggested in one of the UNCOD preparatory papers that it was 'sustained decline in the yield of useful crops from a dry area accompanying certain kinds of environmental change, both natural and induced' (Warren and Maizels 1977, p. 173).

However, when it was first used, by Aubréville (1949), the word desertification referred to a rather different concept. Aubréville saw desertification as an extreme form of 'savannization', the conversion of tropical and sub-tropical forests into savannas. Desertification involved severe soil erosion, changes in soil properties and the invasion of dryland plant species. This sense differs from that established in the UNCOD proceedings in that the latter focused specifically on degradation within drylands, and incorporated new phenomena such as the rise of salinity on irrigation schemes (see Warren and Maizels 1977). Furthermore, definition has been confused by the subsequent coining of rival and overlapping terms such as aridization, aridification, or xerotization (Verstraete 1986).

Several writers have relatively recently started to express reservations about the universal applicability of concepts of overgrazing and desertification (Warren and Maizels 1977, Sandford 1983, Homewood and Rodgers 1984, 1987, Hogg 1987a, Horowitz and Little 1987). Thus Watts (1987) warns against global generalizations based on our persistant ignorance about the Sudan-Sahel region, and argues that many analyses of ecological change are based on problematic assumptions about livestock management or agriculture. Horowitz and Little (1987) stress the difficulty in identifying long-term secular degradation of rangelands, and Homewood and Rodgers (1984, 1987) question the whole concept of overgrazing. The key to their disquiet with the concept is that while almost all the major agencies (FAO, UNESCO, UNEP, World Bank) speak about it, they fail to define the process clearly or to produce evidence of its occurrence. Similarly, Hogg (1987a) argues that 'overstocking' or 'overgrazing' are rarely defined, and that judgements about carrying capacity are subjective, but that subjectivity is rarely admitted. The data necessary to assess 'long-term degradation' of vegetation or desertification simply do not exist. Furthermore, there are real problems of obtaining adequate quantitative data either on the responses of vegetation to different stocking levels or on livestock numbers. Figures of 'carrying capacity' can take no account of seasonal variations in fodder availability or annual variations. They concentrate on absolute numbers of livestock and not densities, and rarely consider spatial mobility. They are therefore of little value in understanding either rangeland ecology or pastoral practice. Although they have become both entrenched and self-reinforcing, they are an unsatisfactory and sometimes dangerous basis for management.

African rangelands are subject to major seasonal variation in productivity and considerable inter-annual variation. While certain crude generalizations about the relations between rainfall and the productivity of herbivores seem to hold true (Coe *et al.* 1976), there are many other factors, notably soil nitrogen and phosphate (Breman and de Wit 1983) and the impact of fire (Norton-Griffiths 1979). Ecological studies tend to be of short duration (often confined to the fieldwork period associated with a Ph.D. thesis) and localized. Yet African grazing ecosystems undergo considerable and important fluctuations from year to year and place to place. Indeed, indigenous pastoral ecosystems seem well-adapted to exploit the spatial and temporal variability in production, adapting herd composition and using movement to maximize survival chances. The Turkana in Kenya, for example, have mixed herds, with camels, which use browse resources which are available even in the dry season, and cattle, which are more productive in the wet season but have to move out of the plains into the hills in the dry season (Coughenour *et al.* 1986). Such systems offer a relatively low output compared to modern capitalist systems such as ranching. However, they are remarkably robust in terms of providing a predictable, if poor, livelihood.

Homewood and Rodgers (1987) argue that the concept of carrying capacity is not suited to areas with great annual variation in primary productivity, and furthermore that it can only be defined satisfactorily in terms of a given management aim. What suits a nomadic pastoralist may not suit a rancher; many African systems have a subsistence stocking rate higher than commercial ranchers would adopt, giving low rates of production per animal but high output per unit area. Furthermore, they suggest that attempts to 'improve' pasture, by creating a more uniform sward of productive and palatable species, may in fact predispose the system to crash under environmental variation. Finally, they argue that there is no one simple ecological succession towards an overgrazed state, but that overgrazing can take many forms, not all of them serious, and it can proceed by diverse routes some of which can be reversed more easily than others, and some of which are more sensitive to particular management than others.

Despite the obvious confusion over what is meant by overgrazing, the concept has been used to justify successive policy interventions in Africa. Horowitz and Little (1987) describe an 'intellectual tradition of anti-nomadism' which has legitimized policies aimed at confining, controlling and often settling nomads. Hogg (1987a) suggests that in the case of northern Kenya the symptoms of desertification are mistaken for the cause. He argues that the pastoralists are increasingly prevented by outside factors from managing their resources in a sustainable way. Large areas of grazing land are virtually deserted while other areas are densely populated and degraded.

Sahelian drought and desertification

For better or worse, the word desertification is now in near-universal use. With time, both popular and academic accounts of desertification have become progressively more sophisticated with regard to causation. Thus recent accounts, for example that by Harrison (1987), give explicit recognition to the subtle and complex interplay of human and environmental factors in environmental degradation in Africa. Climatic variation and the occurrence and persistence of low-rainfall periods in the Sahel, the nature of social and economic activity and the interaction of people with the environment are all important. There are, broadly, three sets of factors discussed: the impact of climatic change, the impact of people degrading their own environment, and the impacts of social and economic relations.

In the last two decades, during which Sahelian desertification has become a global issue, understanding of the nature of climatic variability in the Sahel has grown a great deal. The existence of a series of dry years in tropical Africa beginning at the end of the nineteenth century was identified by the 1950s, but the existence of climatic variations on such a time-scale was little understood by climatologists at least until the 1970s (Lamb 1979). There is now a better picture of climatic variation in Africa within the period of meteorological records, approximately the last 100 years (Tyson *et al.* 1975, Bunting *et al.* 1976, Grove 1981, Dennett *et al.* 1985), and a number of accounts of recent rainfall trends (e.g. Hare 1984). The 1950s were a particularly wet period in the Sudan and Sahel Zones of Africa. They were soon followed by the Sahelian drought which first caught the headlines, which began in about 1968 but was at its peak between 1972 and 1974 (Kowal and Adeoye 1973, Ogoyuntibo and Richards 1977). However, as Lamb (1982) argues, drought did not end in 1974. Hare comments, 'the sense that the Sahelian drought ended in 1974, and with it the need for political concern, was illusory' (Hare 1984, p. 15).

There have been a number of attempts to analyse the periodicity of drought in the Sahel during this century. Tyson *et al.* (1975) used data from 157 stations in South Africa with records from 1910 to 1972 to show that no pattern of progressive desiccation seemed to exist, but did find weaker regional oscillations at intervals of between 2 years and 10-12 and 16-20 years. Winstanley (1973a, b) argued that there was an inverse correlation between rainfall in the Sahel and rainfall in North Africa and the Middle East in winter and spring. He suggested that this might be linked to a strengthening of the northern circumpolar vortex. However, work by Bunting *et al.* (1976) on five West African stations with records of 50 years or more found no clear trends or periodicities, and suggested no linkage with northern hemisphere westerly weather. Faure and Gac (1981), using data on the discharge of the River Senegal, suggested that low-rainfall events had occurred in this century with a periodicity of 31.3 years, and predicted that the Sahelian drought might end in 1985. Clearly, and tragically, it did not.

Drought has remained a persistant feature of tropical Africa, but (if only because more data have become available) its occurrence seems to have become more complex in both space and time. Dennett *et al.* (1985) demonstrate that rainfall decline has not been uniform through the wet season in the Sahel. They suggest in particular that in several years August rainfall in particular has declined by a large amount. August is a critical month with regard to agricultural crop growth, and thus total or partial crop failure can occur without necessarily being revealed in aggregate statistics. Similarly, Hulme (1987) analyses the date of onset and termination of the wet season in the central Sudan. In areas with 200-300 mm of rainfall, wet season failures (seasons too short for crop growth) occurred for the first time this century in the 1970s, and shorter wet seasons and mid- and late-season breaks in the wet season occurred more frequently in drought periods. The variability of tropical climate is underlined by the occurrence of high rainfall in sub-Saharan Africa in 1988.

Climatic changes are not confined to the current century. There is also an increasingly good, if less complete, picture of changes over a longer period (Schove 1977, Nicholson 1978, 1980). Nicholson (1978) suggests that wetter conditions pertained along the Saharan margins throughout the sixteenth to eighteenth centuries, but that a series of major droughts occurred within this period, between 1681 and 1687, between 1738 and 1756 and between 1828 and 1839. In addition there were other, more localized, droughts. By the end of the eighteenth century general conditions were becoming more arid, and dry conditions persisted through the 1828-39 drought until wetter conditions began in the third quarter of the nineteenth century. Precipitation was continuously above the present level between 1875 and 1895 on both margins of the Sahara, in the central Sahara and in much of East and North Africa. This period, of course, was that in which colonial expansion began in much of Africa. The Europeans arrived at the end of a period exceptionally blessed with rainfall.

The variability in Sahelian rainfall in recent centuries can be set against an increasingly detailed understanding of late Quaternary climatic variation in the Tropics over the last 20,000 years. The aridity of the Tropics during the last glacial maximum (Street 1981, Goudie 1983a, Williams 1985) is now well established. Evidence of the former extent of dunes and other landforms south of the Sahara and elsewhere in Africa (Grove 1958, Grove and Warren 1968, Michel 1980, Talbot 1980, Thomas 1984) and of former lake levels (see for example Washburn 1967, Street and Grove 1976, Street-Perrott *et al.* 1985 and Shaw 1985) demonstrates the variability over millennial timescales of prolonged periods of extremely low rainfall in the Sahel and other parts of semi-arid Africa. Similar palaeoenvironmental studies, for example in South America (e.g. Colinvaux 1979, Tricart 1985) and in tropical mountain environments (e.g. Flenley 1979, Hamilton 1982), support the view of variability in past climates in the Tropics, and widespread aridity at the peak of the last glaciation.

Inevitably, greater understanding of past climatic variations has led to attempts to elucidate causation (Lockwood 1986, Flohn and Nicholson 1980). It is now clear that an understanding of the causes of climatic variation in the tropics must embrace the global atmospheric system. The best evidence of this is the demonstration of links between Sahelian rainfall and worldwide sea temperatures (Folland *et al.* 1986), with the implications that this has for the linkage of tropical droughts and atmospheric carbon dioxide, or the 'greenhouse effect'. Of greater interest to those concerned about desertification, however, has been the research on the effect of land surface conditions on rainfall in arid areas, and in particular the possible role of people in causing or exacerbating climatic change through bio-geophysical feedback (Nicholson 1988). Atmospheric variability is driven by both internal and external 'forcing' factors. External factors include earth surface conditions such as soil moisture, vegetation cover, sea surface temperature and the amount of snow and ice. Surface forcing is of particular importance in tropical regions (Nicholson 1988).

Interest in the influence of the land surface on climate has focused on questions of albedo, vegetation cover and surface temperature, and atmospheric dust. All these have been influenced to some extent by human impact on a global scale, but although total impacts are relatively trivial (Henderson-Sellars and Gornitz 1984), local effects appear to have been of considerably greater magnitude. Gornitz uses a variety of data sources to attempt to quantify anthropogenic vegetation changes in West Africa over the last 100 years through grazing, logging and cultivation (Gornitz and NASA 1985). He suggests that a 56 per cent loss of forest may have occurred in the whole of West Africa since 1890, with higher rates of loss in the Ivory Coast, Ghana and Liberia of between 64 and 70 per cent. Estimates of total land cover conversion in West Africa lie between 88 million hectares (using a generated map of land use intensity) to 123 million hectares (using extrapolated forestry data). This implies an increase in albedo of 0.4 per cent, compared with global figures of between 0.00033 and 0.0064 per cent (Gornitz and NASA 1985). Gornitz suggests that the total increase in albedo since agriculture began might be of the order of 0.5 per cent, thus four fifths of present albedo levels may be due to earth surface changes within the last century.

Other studies within the Sahel have identified seasonal and inter-annual variations in albedo on a small scale (Courel *et al.* 1984), while satellite imagery from the Advanced Very High Resolution Radiometer on the NOAA polar orbiting platforms demonstrates an apparent decrease in vegetation cover between 1981 and 1984 (Tucker *et al.* 1985, 1986). There have also been studies of the occurrence of atmospheric dust in the Sahara in relation to drought (Prospero and Nees 1977, Middleton 1985). One of the various hypotheses linking dust to rainfall suppression links it to the creation of warm inversion layers at altitude which suppress convective rising of humid surface air. Photography from the US Space Shuttle missions provide some support for this hypothesis.

Links between albedo changes and climate were drawn by Otterman (1974), who reported a temperature difference of 5 °C between the heavily grazed Sinai desert and the less grazed Negev. He suggested that the baring of soils through overgrazing might be a possible mechanism of desertification, in which higher albedo would cause lower surface temperatures and reduced atmospheric convection. These observation were subsequently backed up by modelling experiments (e.g. Charney 1975, Charney *et al.* 1975) which produced interesting results, and some controversy (Idso 1977 and the debate in *Science* volumes 189 and 191). The notion that, in Charney's memorable phrase, 'the desert feeds on itself' is now widely taken as fact. It is better viewed as a hypothesis because the evidence for it remains inconclusive. For example, Charney's model was based on the investigation of an albedo change from 0.14 to 0.35, vastly different figures from those Gornitz and NASA (1985) discuss. There is clearly much room for both speculation and further research. Wendler and Eaton (1983) argue, for example, on the basis of studies in the pre-Saharan steppe of Tunisia that the albedo of desert vegetation is in fact higher, and the increase in albedo when vegetation is grazed intensively is therefore less than modelling studies assume.

A further problem with these studies is that the effect of vegetation on albedo cannot easily be separated from its other effects, for example soil moisture (Nicholson 1988). A study in the Sonoran Desert (Jackson and Idso 1975) found bare land to be warmer than a vegetated surface, in contrast to Otterman's results (1974). However, vegetation types in the two areas are not directly comparable (Nicholson 1988), and the controversy serves to demonstrate the difficulty of producing clear and simple conclusions from a small number of empirical (and often remotely-sensed) observations and powerful computer models.

Clearly surface conditions are an important input into climate dynamics in the tropics, and human action does influence those conditions. Locally at least, that influence can be significant. However, the linkage between human agency and climatic variation remains, for the present, in need of convincing scientific demonstration. Nonetheless, phenomena of desiccation and degradation certainly exist, and the role of human action in creating desert-like conditions remains an important subject both of research and popular concern.

Ideology and desertification

Concern about environmental degradation in arid and semi-arid areas is not new. Over time, both the focus of attention and the kinds of explanations canvassed have shifted, moving in the case of the Sahel for example from arguments about the southwards invasion of desert conditions into new areas to more recent ideas of degradation of semi-arid areas from within. The remarkable features of desertification consciousness, however, lie not in the way it has changed, but its persistence, continuity and geographical spread. In areas such as West Africa it is possible to trace a continuous thread of environmental concern in colonial and post-colonial states, and latterly in the international arena, from early in this century to the present day. The nature of this West African experience reveals a great deal about the ideology of environmentalism, and hence of thinking and policy aimed at non-degrading or sustainable development.

The first focus of concern in West Africa concerned the creation of desert conditions by the physical advance of the desert into more moist areas, a process of desiccation. The fact of desiccation in Africa over geological time was well-established (e.g. Hobley 1914), but that was not the point at issue here. Rather, it was the fear of increasing aridity over the short term, and specifically the encroachment southwards of the Sahara (e.g. Bovill 1921). Droughts in the early decades of colonial rule (for example in Hausaland in 1913, Grove 1973) appear to have had significant effects on the thinking of colonial administrators. For example, low rainfall and low river floods in Sokoto Province in Northern Nigeria in 1917 and 1918 suggested to the resident in Sokoto that desiccation was taking place. This concern led to a remarkable episode of interest in small-scale irrigation using floodplain bunds in the valley of the River Sokoto between 1919 and 1921 (Adams 1987). Although these experiments were fairly unsuccessful, concern about desiccation persisted and spread. For example, the problem was mentioned in 1921 in Lugard's treatise on colonial policy *The Dual Mandate*.

When the forester E. P. Stebbing visited West Africa in 1934, a representative of the Emir of Katsina was among those who showed him evidence of recently deteriorated conditions in the far north of Nigeria. He suggested that West Africa was undergoing progressive desiccation and that the Sahara was moving southwards: the 'silent invasion of the great desert' (Stebbing 1935, p. 518). He saw the savanna as a form of open deciduous forest, progressively degraded by burning and shifting cultivation, grazing and browsing and then destructive pollarding by graziers: 'Under this system of treatment the final extinction of the savanna forest takes place, when the weakened roots and vanishing rainfall result in the death of the trees' (Stebbing 1935, p. 513). These arguments were substantively refuted by Jones (1938) on the strength of the fieldwork of the Anglo-French Forestry Commission to northern Nigeria and Niger in 1936 and 1937. He argued that there was no evidence of southward encroachment of sand, retrogression of vegetation, permanent reduction of rainfall, the recent shrinkage of streams or lakes, or the lowering of water table. He concluded 'there is no reason to fear that the desiccation through climatic causes will impair the habitability of the West African colonies for many generations to come' (Jones 1938, p. 421).

Subsequently, Stebbing went some way towards this position, stressing instead the spoilation and erosion of land due to population increase and to agricultural and pastoral practices (Stebbing 1938). The debate was reviewed by Stamp (1940), who concluded that 'there is no need to fear desiccation through climatic causes'. On the other hand, 'the spread of man-made desert from within is quite another matter' (Stamp 1940, p. 300). Concern thus shifted to processes of in-situ deterioration involving soil erosion. This was obviously affected by the global dissemination of the experience of the American Dust Bowl. Soil erosion and dust storms in the American Mid-West, and particularly the experience of dust over the eastern seaboard, generated huge public concern in the USA, with concomitant theoretical and institutional responses regarding the position of ecology and the establishment of the United States Bureau of Reclamation (Worster 1979, 1985). Perception of this environmental crisis in the USA passed via newspapers and other routes (including the Imperial School of Tropical Agriculture in Trinidad, Anderson 1984, Anderson and Millington 1987) to inform and feed concern about soil erosion in arid and semi-arid environments elsewhere. These took root most obviously in Africa (see Beinart 1984; Anderson 1984, see for example Harroy 1949; Aubréville 1949), but the scope of concern was essentially worldwide (Jacks and Whyte 1938, Furon 1947).

In West Africa, Stamp suggested, 'There now seems little doubt that the problem before West Africa is not the special one of Saharan encroachment but the universal one of man-induced soil erosion' (Stamp 1940, p. 300). Similarly, Vogt concluded: 'Whether or not Africa is actually suffering a climatic change, man is most effectively helping to desiccate the continent' (Vogt 1949, p. 248). Although subsequent reviews suggested no further deterioration in environmental conditions in northern Nigeria, concern certainly persisted locally in Africa. In northern Nigeria it led to an application for money from the United Nations Special Fund for a study of land and water resources in the Sokoto-Rima basin (Adams 1987). That study recommended irrigation development on a large scale. Ironically, the Bakolori Project, developed in the Sokoto Valley in the 1970s, has itself adversely affected existing indigenous irrigation in the floodplain of the river (Adams 1985a, b).

Clearly it is to an extent true that desertification is in the eye of the beholder; historically people have seen what they wanted to see. Preconceptions about ecological processes, an exaggerated ability to infer process from form, and therefore make assumptions about what is happening in a landscape on the basis of what can be seen at any given moment, and above all the ideology of the observer all contribute to the perception of desertification. The importance of ideology in affecting environmental perception is well demonstrated by Anderson (1984) in the context of colonial East Africa. He describes four factors which encouraged a move by the colonial states to undertake direct intervention in African agriculture over the 1930s: the Depression, the American Dust Bowl, population growth and drought.

The Depression threatened both settler and African farmers. African farmers responded to the slump in agricultural prices by expanding the area under production, of cotton and coffee in Uganda and maize in Kenya and Tanganyika. White farmers facing bankruptcy responded defensively, seeking to entrench their position, in the process claiming (notably to the Kenya Land Commission) that African farming practices were damaging to the environment and unproductive. The impacts of the Depression and the political response to it were compounded by images of the American Dust Bowl which 'caused Agricultural Officers all over British Africa to examine their own localities for signs of this menace' (Anderson 1984, p. 327). The prevention of erosion, particularly as this was perceived to occur primarily on African land, was a major stimulus to intervention in African agriculture. Meanwhile, by the 1930s the growth of the African population had become a matter of concern, partly because of the problems of overcrowding and landlessness in the Kikuyu reserves (Throup 1987). European settlers farming in the White Highlands perceived soil erosion both on the reserves and by squatters on their own farms. Finally, the years from the mid-1920s to the mid-1930s were dry in East Africa, and drought created periodic local food shortages and sometimes extensive deaths of cattle. Famine relief by the state was expensive, and drought heightened the perception of environmental degradation caused by African husbandry.

It is not perhaps surprising to find that by 1938 soil conservation had become a major concern of government in the East African colonies, the cutting edge of a policy of state intervention in African agriculture. The details of the implementation of policy are diverse, but what is most important here is the complexity of the context in which they emerged. Political and economic concerns of white settlers were combined with the occurrence of natural phenomena (drought), wider economic concerns (the Depression and the growth of African agriculture) and the existence of a powerful and emotive image of environmental decline in semi-arid regions under the pressure of agricultural misuse. The result was a persuasive and self-confirming perception of environmental degradation, or what would now be called desertification. Looking back at the records, there are few data with which to assess scientifically the severity of any such problems. Such data were not collected. They were not necessary; the ideology of degradation was sufficient to generate policy.

Much the same pattern persists today. Despite the apparent power of ecological science to sidestep partial knowledge and offer a definitive statement about environmental change, and despite their faith in their science, ecologists of recent decades have suffered from similar problems to those afflicting their colonial predecessors. Data are without doubt too few to refute or sustain the hypothesis that serious desertification is taking place on any scale, but that does not justify the dominant position of the ideology of man-induced environmental decline. The key to understanding its power lies in the strength of neo-Malthusian thinking within environmentalism. This is present in muted form in the *World Conservation Strategy*. Here desertification is seen to be caused primarily by human transgression of ecologically-defined limits. Although there are problems of 'unwise development projects' and the 'inadequate management' of irrigated areas, the key problem is seen to be 'pressure of human numbers and numbers of livestock' (IUCN 1980, para 16.9).

The neo-Malthusian view is readily found in starker form in writings on desertification, particularly in studies of Africa. One example is Curry-Lindahl, who suggests that 'for decades past, ecologists have been warning us about the serious environmental consequences and ecological undermining of present land-use practices' in semi-arid lands: 'oversized stocks of cattle, goats and sheep, as well as unwise agricultural methods, coupled with extensive collecting of firewood and excessive burning, had long been paving the way for desertification and 'death' of productive lands' (Curry-Lindahl 1986, p. 106). The Sahelian drought of the 1970s focused attention on the problem, but governments and aid agencies drew the wrong conclusions: 'They called the effects of the drought a "natural disaster" and blamed the climate! The real root of the problem, that Man himself was responsible, was not recognized or admitted: it was too unpopular a message' (p. 107).

However, it was a version of that message which was presented at the UN Conference on Desertification, and it is a further softened and modified version which informs the WCS. Its roots are in neo-Malthusian thinking. Thus as long ago as 1949 Vogt wrote:

> The European in Africa has temporarily removed the Malthusian checks. He has put down tribal wars, destroyed predators, moved enough food about the continent to check famine - but he has not substituted constructive measures to balance his destruction of the old order.
>
> (Vogt 1949, p. 260)

Vogt's central concern is carrying capacity, and he argues that 'Man has moved into an untenable position by protracted and wholesale violation of certain natural laws' (p. 264). This idea has persisted in thinking about desertification. For example, Curry-Lindahl argues that:

> the fact that each area has a carrying capacity beyond which it cannot be utilized without causing damage, deterioration, and decreased productivity, is an ecological rule that is almost always forgotten by those who plan and carry out land-use development schemes.
>
> (Curry-Lindahl 1986, p. 125)

The ways in which these neo-Malthusian ideas relate to thinking in ecology has been described in Chapter 2. In discussing desertification an essentially ecological framework for planning is made explicit, and becomes fully incorporated into sustainable development thinking. The ecological approach to the degradation of arid drylands first developed in the context of the American Dust Bowl in the 1930s (Worster 1979, 1985). The report of the Great Plains Committee in 1936 ushered in 'a radically new environmental outlook' (Worster 1985 p. 232), one based on the ecological thinking of men like Clements, whose notion of climatic climax, Worster argues, came to dominate the national imagination as well as ecological thinking. The functional anthropocentric possiblism of the sodbuster farmers was replaced by 'a more inclusive, coordinated, ecological perspective', both more communal and less individualistic (Worster 1985, p. 232). Although this holistic view was soon challenged by the rise of the reductionist thinking that has dominated ecology since the 1940s, it remains an important strand in thinking about desertification in the Tropics. Furthermore, it proves a powerful basis for policy-making. The notion that human and livestock population expand to exceed the carrying capacity of semi-arid regions remains entrenched for example in tropical Africa. Thus presumed overstocking of African cattle in Baringo District in Kenya led to the suggestion of culling through cattle tax or compulsory purchase, and 'carrying capacities' were calculated and targets established (Anderson 1984), despite the unpopularity of such campaigns elsewhere (Tignor 1971).

The political economy of desertification

Ideologies of desertification have changed over time. Environmentalists who have learned enough human ecology to reject simple climatic determinism rightly identify human agency as a key factor in the nexus of poverty and degradation in the Sahel. Yet their analysis remains constrained by a very limited view of social relations. In the

memorable words of Watts, social perspectives on drought and desertification too often degenerate into 'a pluralist grab-bag of ideas embracing everything from land tenure to international political organisation' (Watts 1987, p. 188).

Neo-Malthusian blinkers do not help the environmentalist make sense of the Sahel. Thus Sinclair and Fryxell (1985) identify two contrasting views of the crisis in the Sahel, and particularly that of Ethiopia in 1984. They suggest that the dominant view, the 'drought hypothesis', which suggests problems are due to the failure of the rains exacerbated by warfare, is misconceived. Instead they suggest a 'settlement-overgrazing hypothesis'. They argue that until about thirty years ago the 'normal' land use pattern in the Sahel was based on migratory grazing using seasonally available resources. They suggest that this system had been operating in a 'balanced and reasonably stable' way for many centuries, possibly since domestic cattle first appeared in the Sahel 5000 years ago. It broke down 'through well-intended but short-sighted and misinformed intervention through aid projects'. Problems began after the Second World War exacerbated by population growth, overgrazing, and agricultural practices aimed at short-term profit not sustained yield (Sinclair and Fryxell 1985, p. 992). Sinclair and Fryxell suggest that overgrazing brought about the 'regression of plants' around boreholes and wells, and as these 'piospheres' (Warren and Maizels 1977) expanded and joined up, extensive areas of the Sahel became desertified. They then present the arguments about the possible feedback effects of bare desertified soil on climate, and suggest that the Sahelian ecosystem 'is being pushed into a new stable state of self-perpetuating drought' (p. 992). Man degrades the land, and in their analysis the fault lies with the aid agencies who fund projects which break down the older and sustainable migratory pattern. In particular, in their stark Malthusian analysis, short-term food aid by itself will 'only make the situation worse', since 'simply feeding the people and leaving them on the degraded land will maintain and exacerbate the imbalance and not allow the land to recover' (p. 992).

Sinclair and Fryxell (1985) certainly highlight an important aspect of environmental degradation in the role of development agencies, but they do so in complete isolation from wider debates about political economy, and specifically about the relations between political economy and environment. There is now a growing literature which makes an explicit attempt to elucidate these relations. Blaikie (1985) argues that soil degradation and erosion are the result of sets of decisions about land use made over time by land users, and that these cannot be isolated from their political economic context. In the words of Jean Copans 'we can no longer separate the natural phenomenon and the necessary political translation of its effects' (Copans 1983, p. 94). The focus of this work has been the impact of colonialism, although as

Blaikie and Brookfield point out, the damaging phenomena of that experience, 'taxation, land-grabbing or amassment, demands for cash crop production and the extraction of labour and surplus' (Blaikie and Brookfield, p. 121) are not unique to colonialism: they are features of the expanding capitalist world-economy which have been experienced more or less everywhere at different periods and to different degrees.

The political economy approach provides a drastically different interpretation of environmental degradation in the Sahel, and elsewhere in semi-arid Africa. Radical interpretations of the impact of imperial trade, colonialism and dependent economic relations on the social and economic relations of Third World societies, for example Kenya (Leys 1975), Niger (Roberts 1981), or Nigeria (Forrest 1981, Shenton 1986), follow relatively well-defined routes. Relatively few draw explicit and effective links between international political economy and the peasant household. Copans (1983), writing about the social and political context of the 1971-6 drought in the Sahel, argues that the phenomenon of famine must be understood in the context of a series of scales: first, the conditions of production; second, the social organization of production; third, the national political economy (inequalities of both class and region); and fourth the international political economy (dependency and neocolonialism). He is concerned to demonstrate that real explanations of rural poverty and exposure to famine often lie outside the rural community itself, just as they lie outside ecology and climatology. Understanding of the dynamics of poverty and environmental degradation must be based in part at least on consideration of the economy and ecology of peasant production.

The impact of the history of social relations in the colonial Third World is a lot better known than the impact this has had on the natural environment. What is needed therefore is a new political economy, embracing the environmental dimension, of the kind Corbridge (1986) argues for. Copans suggests that Africanists failed to understand the economic and political causes of the 'actual ecological imbalances' which occurred during the 1970s Sahel drought (Copans 1983, p. 83). He criticizes the research 'boom' which followed the drought for its conventional nature and disciplinary constraints, and highlights the lack of work on the relations between social, economic and political history in Africa and the history of the changes in the natural environment which these produced. He therefore suggests a new focus on the impact of modernization on both the social and natural environment.

Several authors achieve this, notably Watts (1983a, b, 1984, 1987), in his studies of the village of Kaita in Katsina Emirate in northern Nigeria, and Franke and Chasin in their book *Seeds of Famine* (1980). Franke and Chasin begin their account of the famine in Sahelian countries in the early 1970s with a description of pre-colonial West Africa. They suggest that over time societies had adapted to the deferential environmental zones of the Sahel and created 'relatively prosperous and stable communities' (Franke and Chasin 1980, p. 61). Despite wars and the exactions of elites they see a balance in favour of 'environmental preservation' (p. 62). This picture changed in the seventeenth century, largely with the expansion of European economic power and trading influence, and latterly, at the end of the nineteenth century, by colonial expansion.

Their story then focuses on the role of the French colonial governments in Afrique Oriental Français in bringing about the expansion of groundnut (peanut) cultivation through a mixture of incentive and coercion (particularly taxation and forced labour), and the commoditization and monetization of the economy. Regular shipments of groundnuts to Marseilles began in 1884, and they were exempted from duty in 1892. Railways and roads, and after 1913 agricultural research, credit and extension, were all marshalled to promote groundnut production. Production rose through the colonial period and after independence. Production in Senegal was 45,000 tons in 1884-5, 0.6 million tons in 1936-7 and over 1 million tons in 1965-6. By 1964 groundnuts represented 58 per cent of exports by value in Mauretania, 59 per cent in Mali, 63 per cent in Niger and a staggering 79 per cent in Senegal.

Franke and Chasin argue that individual farmers were caught in a production trap with groundnut yields declining, demanding the expansion of production into fallow areas and into areas used for growing staple food crops of millet. As a result, cash was required to purchase foodgrains, demanding further groundnut production, while soil exhaustion caused further declines in yields. A similar intensification trap worked at the level of national economies. Declining terms of trade told against these one-crop economies, while competition in the post-war period for the European market from other oilseed producers, particularly American soya oil, reduced prices. The expansion of groundnut production became a national goal, with production doubling in Senegal for example between 1954 and 1957, but it carried serious implications of soil exhaustion and the erosion of food self-sufficiency. It was this which created conditions in which Sahelian producers were exposed to famine during the drought of the early 1970s.

In his work in Nigeria, Watts seeks to demonstrate that social relations, that is the relations between producers and non-producers and the social mechanisms for surplus extraction, are the means of understanding the 'connections between material circumstances and ecological conditions' (Watts 1987, p. 190). His analysis starts therefore from the peasant household, locked into the local economy and ecology. He discusses the impact of capitalism on production systems in rural Nigeria, looking in particular at the articulation of the pre-capitalist mode of production with the global capitalist system through the agency of the colonial state (Watts 1983b). With colonial rule came taxation, cash cropping (in the north of Nigeria chiefly cotton and groundnuts) and railways.

The railway reached Kano in 1913 and generated a rapid rise in groundnut exports and (one must assume) a parallel shift in land use. While some may have profited from this trade, it is argued that these changes reduced the 'margin of security' which poorer Hausa farmers had previously enjoyed (Watts 1983b, p. 251), both from the highly adaptive ecology of their production system (based on intercropping and the skilled exploitation of the diversity of upland and wetland environments open to them) and the range of sources of livelihood outside farming (crafts, farm labour, livestock, seasonal migration and sale of land).

Hausa farmers became progressively more involved in cash cropping. Falls in groundnut prices produced a 'reproduction squeeze' where farmers either increased production or reduced consumption. This squeeze was felt unequally by the poor in the differentiated Hausa society of northern Nigeria, and it promoted (among other things) the decline of the moral economy, the patterns of reciprocity between households, particularly between rich and poor, which had provided a safety net in times of drought or disease. Poor farmers, 'shackled by their own poverty, are largely powerless to effect the sorts of changes which might mitigate the debilitating consequences of environmental hazards' (Watts 1983b, p. 256). With drought, in this century as before, came famine (Apeldoorn 1980). Watts sees drought 'refracted through the prism of community inequality' (p. 256), and the environmental crisis of drought something which reveals the structure of the social system reworked by colonialism and maintained by international political economy.

Frank and Chasin (1980) develop a slightly different argument about groundnut production, stressing the unsuitability of soils in the Sahel and Sudan zones to sustain continuous production. They argue that before cash cropping became dominant, an 'ecologically sound' system of crop production was practised, rotating groundnuts and millet, and following three years of groundnut production with a fallow period of six years to allow phosphorus, magnesium, potassium and organic matter to recover in the soil. Franke and Chasin (1979, 1980) claim that continuous sole-cropping with groundnuts and the violation of fallow led inexorably to soil deterioration in agricultural regions. The expansion of groundnut cultivation northwards, and the cultivation of fallow lands further south in turn disrupted the pastoral economy, creating conflicts over dry season pastures and additional pressure on seasonal northern grasslands. Thus cash-cropping, driven initially by colonial policy and latterly by the demands of the national economy and by international political economy, created both degraded croplands and desertified rangelands in West Africa. In this lay the seeds of famine.

The sweep of this argument is undoubtedly attractive, although its history is rather simplistic. There is for example no analysis of factions within either the colonial state or the African producers, no consideration of winners and losers within the fundamental system. Nonetheless, its basis, which reflects widely-held views about the nature of the colonial impact in Africa, is probably basically sound. The details of the ecological implications of continuous agricultural production do however remain to be spelled out, just as the nature of change in pastoral and agricultural ecosystems need to be investigated through detailed case studies. Nonetheless, the existence of important links between political economy and environment is clear. The impacts of the actions of colonial and post-colonial states on the actions of farmers and pastoralists, and the effects of their actions in turn on the ground must be the centrepiece of accounts of environmental degradation.

Sustainability and desertification policy responses

The nature of ideologies of desertification of course influences understanding of problems and the nature of policies recommended. The 'overgrazing hypothesis', linking land management and climatic change through bio-geophysical feedback loops, readily becomes the foundation stone of policy. Thus Hare reviews the work done in modelling the impacts of albedo, soil moisture and dust on climate, and concludes that desertification 'harshens the microclimate', thus making more difficult 'the reestablishment of biological productivity' (Hare 1984, p. 19). He suggests that the best defence against this harshening is land use control, meaning fencing and the control of the movement of animals. Thus measures to control desertification not only 'restore plant cover and soil conditions', but 'repair the microclimate' (p. 19).

Sinclair and Fryxell (1985) offer a series of proposals for action in the Sahel based on similar premises, although they are less concerned with preventing the movement of livestock than with the recreation of free migratory movements. Their proposals have a similar focus on control. The regeneration of degraded environments can only take place if the land is 'rested' until vegetation can recover. They suggest the following strategies:

- move people from the degraded land to new areas,
- educate and help people in those new areas to establish a new and suitable rural economy,

- institute education and family planning to control population growth,
- severely restrict and cull cattle herds,
- monitor vegetation succession on degraded land,
- when recovered establish a modified migration or rotational grazing system,
- only construct wells if they do not harm the migration system,
- encourage African governments to institute these measures.

These proposals are all predicated on the ending of 'uncoordinated and ecologically damaging aid-giving' by international donors whom Sinclair and Fryxell see as the main culprits in the degradation of the Sahel. The problems of the area can therefore only be found in drastically altering that developmental intervention. Similar arguments were put by Ormerod. He suggested that economic demand, and attempts by international organizations to satisfy it by increasing local production, was a major factor in increasing agricultural degradation in tropical areas. He argued that the eradication of the tsetse fly and removal of other barriers to trade and development were problematic, and concluded that 'there is a particular danger in stimulating economic advance in desert areas, and in particular in linking nomads to the monetary system' (Ormerod 1978, p. 377).

Unfortunately, neither of these analyses generate feasible and humane directions for sustainable development. Ormerod's view fails to take into account the fact of the penetration of capitalism and the extent of monetization, and offers limited prospects of improved welfare for desert margin people. The policies suggested by Sinclair and Fryxell are little stronger, being naïve in both a practical and political sense. They place excessive faith in the power of population control and family planning programmes, or at least those which stop short of coercion, to affect the rate of population growth. They suggest simply the 'establishment of 'a new and suitable rural economy', without apparently appreciating the difficulties of even conceiving of such a thing, let alone the politics and economics of achieving it. They suffer from the blindness of many commentators on the presumed ills of African grazing systems by assuming that culling, even if deemed necessary, is feasible. Experience elsewhere in Africa makes it quite clear that such state initiatives can be both misconceived and hugely unpopular. They also tend to fail (Tignor 1971).

Above all, Sinclair and Fryxell appear to fail to appreciate the problems of induced population resettlement in the Third World, either the practical difficulties or the associated human costs involved. The problems of population resettlement are discussed in detail in Chapter 7. It is clear that even the relatively limited resettlement exercises connected with reservoir construction in the Third World have a poor record, with initial apparent success often followed by subsequent abandonment, disillusionment and deprivation (e.g. Chambers 1970, Mabogunje 1973, Adams 1985c). The attempt by the Ethiopian Government to resettle people from the drought, war and famine stricken highlands has been the target of humanitarian and political complaint by donor nations. The political and human face of extensive involuntary government resettlement in the Sahel is unlikely to be acceptable.

Similarly, the record of attempts to settle and cultivate virgin areas in Africa is also extremely poor, for a combination of ecological, social and economic reasons (Chambers 1969). The classic failure of this kind is probably the near-legendary post-war Groundnut Scheme in Tanganyika (Frankel 1953, Seabrook 1957), but there are other examples, for example the Niger Agricultural Project in Nigeria (Baldwin 1957), the attempt to establish farm settlements in Western Nigeria on the Israeli *Moshavi* model (Roider 1971), and the ambitious Zande Scheme in the Sudan (Reining 1966). More recent endeavours, for example the Bura Irrigation Settlement Project in Kenya, which involved irrigated cotton production by settlers from all over Kenya in dry Sahelian bush grazed by Orma pastoral people, have been no more successful (Hughes 1985).

These kinds of 'solutions' to the twin problems of poverty and environmental degradation in semi-arid lands are, however, far from unusual. The possibility of moving forwards in marginal areas through the strengthening of pastoral institutions, a much improved version of Ormerod's (1978) vision, was discussed in the papers presented to the 1976 UN Conference on Desertification (United Nations 1977, p. 51). It has subsequently been widely canvassed and indeed implemented, for example by non-governmental organizations such as OXFAM in northern Kenya, but it remains an unusual (if highly promising) strategy. Repeatedly, state intervention in rangelands, often triggered or legitimized by concerns about desertification, has involved attempts to make nomads settle. Thus for example there have been a series of attempts to settle destitute pastoralists in northern Kenya in the 1970s and 1980s, often on irrigation schemes. These have been expensive and, to the pastoralists, unwelcome (Hogg 1983, 1987a, b). They have also been conspicuously unsuccessful. Turkana District covers some 62,000 square kilometres, with erratic annual rainfall of 150-350 mm. The Turkana are pastoralists, with mixed herds of small stock (sheep, goats), camels and cattle. Following a series of droughts since the mid-1960s, famine relief efforts began in Turkana. In 1982 the Turkana Rehabilitation Programme was supplying food to 80,000 people, as a result of which the intense drought of 1984 (53 mm rain in Lodwar, the District headquarters) was ridden out (M.E. Adams 1986).

As part of past attempts to 'merge relief and development' (M.E. Adams 1986), substantial numbers of Turkana pastoralists have been settled along the seasonal rivers of the Kerio and Turkwel in irrigation schemes, and on the shores of Lake Turkana as fishermen (Hogg 1987b). The first irrigation schemes were begun in 1962 by mission organizations as an alternative to pastoralism for Turkana on famine relief. The Kenyan government and a series of bilateral and multilateral aid agencies have subsequently been involved in the management of the schemes, which have proved inordinately expensive. A Ministry of Agriculture appraisal in 1984 calculated that the total development cost of the Turkana cluster of small scale irrigation schemes (some $63,140 at 1983 prices) represented $21,000 per household. This is expensive even by the low standards of African irrigation, and represents fifteen times the cost of setting up each family with a herd of replacement livestock, and the equivalent to famine relief for 200 years (Asmon *et al.* 1984).

Hogg argues that irrigation development in Turkana, and elsewhere, for example Isiolo, was seen as an deliberate element in a strategy against what the government perceived to be the destructive effects of pastoralism, part of the anti-nomadism discussed by Horowitz and Little (1987). Perhaps unsurprisingly, it makes no sense in terms of the normal socio-economic objectives of development, but more seriously, it has failed in its supposed effects on both pastoralists and their environment. Hogg argues that irrigation has contributed to the further marginalization of already poor pastoralists and significantly increased pressure on key grazing lands (Hogg 1987a). Partial sedentarization, combined with increased stock theft in the hills on the Ugandan border to which cattle have to move in the dry season, has caused what Hogg calls 'green desertification' (p. 51) because of under-use, while other areas are degraded through over-use. Clearance of riverine forest vegetation near settlements is probably the most obvious aspect of what Hogg sees as policy-induced desertification. He comments 'sedentarisation has meant a declining resource base and increased insecurity; and desertification, as a result of population and livestock concentration, has continued unabated and largely unchecked' (p. 57). The root causes of desertification in Turkana in Hogg's view then are first poverty, second induced settlement and third modernization. Development itself is therefore implicated in the causation of environmental degradation, and 'desertification control' that does not take on board the need to revise development policies across the board is unlikely to succeed.

In the light of this, it is interesting to consider the nature and success of the anti-desertification initiatives which followed the 1977 UN Conference on Desertification (UNCOD) in Nairobi. A Plan of Action to Combat Desertification (PACD) was adopted at the

Conference which included recommendations for action at national, regional and international scales. The Plan distinguished between immediate and ultimate goals. Its immediate goal was 'to prevent and to arrest the advance of desertification, and, where possible, to reclaim desertified land for productive use' (Walls 1984). The ultimate objective was to sustain or enhance the productivity of arid, semi-arid and sub-humid areas and improve the quality of life of their people. UNEP was made responsible by the UN General Assembly in 1977 for coordinating implementation of the plan (Karrar 1984). Their task was to stimulate regional and national action, to assist in project design, to suggest strategies for project finance, and to promote training and research. Regional action was to consist of pursuit of transnational projects discussed at the Conference (those on a 'transnational green belt in North Africa' and 'management of regional aquifers in North-East Africa' were implemented, Karrar 1984). However, the cutting edge of implementation was to be action by national governments. The PACD required them to:

- establish or designate a government authority to combat desertification,
- assess desertification problems at national, provincial or sub-provincial levels,
- establish national priorities for actions against desertification,
- prepare a national plan of action against desertification,
- select priorities for national action,
- prepare and submit requests for international support for specific activities,
- implement the national plans of action

UNEP established a desertification branch, an Interagency Working Group and a Consultative Group on Desertification Control, and began publication of the Desertification Control Bulletin. Other initiatives, such as the UNEP/MAB Integrated Project on Arid Lands (IPAL) based at Marsabit in Kenya, were continued. The UN General Assembly established (despite opposition from certain donor nations) a Special Account for anti-desertification project finance. However, despite this flurry of activity, implementation flagged. Although UNEP spent over $20 million on desertification between 1974 and 1983, this was but a drop in the ocean. In six years the UN Special Account attracted only $48,500. Only two countries had drawn up anti-desertification plans, Sudan and Afghanistan, although there were nine others in draft (Walls 1984). Few countries had designated agencies for desertification control. Assessments of desertification in 1983 showed it to be increasing not diminishing (Mabbutt 1984, Berry 1984, Tolba 1986). In the Sudano-Sahelian region for example, 90 per cent of rangelands and 80 per cent of croplands were thought to be 'at least moderately desertified' (Mabbutt 1984, p. 106). Reviews of the implementation of the PACD reveal substantial underfunding and lack of interest on the part of governments. At the UNEP Governing Council in 1980 the Executive Director highlighted lack of cooperation within the UN system, inadequate external funding of projects, and the low priority assigned to desertification control by afflicted countries. Beneath these lay basic problems of lack of money and lack of political will (Walls 1984).

UNEP estimated in 1980 that about $90 billion would be required to finance a programme to meet the demands of the core of the PACD over 20 years: $4.5 billion per year. This would have rehabilitated all desertified irrigated land, 70 per cent of rainfed cropland and 50 per cent of rangelands. In fact it was estimated that only $7 billion had been spent combating desertification between 1978 and 1983 ($1.17 million per year, or 0.2 per cent of that needed). Of that $7 billion, only about $400,000 was actually spent on desertification control, the rest going on infrastructural projects such as roads, water supplies, buildings, research and training (Walls 1984, Dregne 1984). While such infrastructure might be necessary, it appeared to have little to do with desertification control, and appeared to result in existing projects in semi-arid areas being relabelled as 'desertification' projects, either by aid agencies eager to show they were doing something or by governments hoping to make projects more attractive by jumping on the latest bandwagon.

The inadequacy of the anti-desertification endeavour in the late 1970s is perhaps the more surprising given the enormous scale of aid agency interventions in Sahelian countries, and the media attention devoted to the question of drought. In 1973 seven Sahelian nations established the Comité Interétats de Lutte contre la Sécheresse Sahélienne (CILSS). Shortly afterwards the UN Special Sahelian Office met with donor nations and Sahelian countries. Promises of aid were made in 1973, and increased in subsequent years. The US Congress supported the notion of a long-term development programme in the Sahel in December 1973 under the Foreign Assistant Act. This is sometimes termed the New Directions congressional mandate because it emphazised redistribution as well as growth, appropriate technology, participation in decision-making and basic needs (Derman 1984). Political changes in Africa (the independence of Portuguese colonies and the revolution in Ethiopia) may have accelerated the tempo of aid-giving. In 1976 USAID began to prepare a report for Congress for action, and in Senegal Henry Kissinger made a public commitment of $7.5 billion to 'roll back the desert' (Franke and Chasin 1980, p. 137). In 1976 an organization, the Club du Sahel, was formed to coordinate aid giving, jointly sponsored by CILSS and certain aid donors.

The result of these initiatives, initially, was a vast research output on behalf of both bilateral and multilateral aid donors, much of it done by university researchers in Europe or (especially) America as consultants. In 1977 the OECD (Organization for Economic Cooperation and Development) published the draft of a strategy and programme for drought control and development in the Sahel. This was adopted by CILSS and the Club du Sahel in the same year (Derman 1984). Its aim was to achieve food self-sufficiency by raising agricultural production and cutting imports. It envisaged projects in three stages: first (1978-82) projects to alleviate drought and reestablish food production and 'ecological recovery' (Franke and Chasin 1980, p. 148); second (1982-90) medium-range projects and third (1990-2000) projects to achieve complete food self-sufficiency. The programme would require massive increases in the production of wheat (seventyfold), rice (fivefold), cattle (twofold) and fish (twofold) on 1970 figures. In large part this was to be achieved by improved rainfed farming (doubling production by the end of the century, working on drought-tolerant crops and off-farm employment), but a massive fivefold increase in the area irrigated (to 1.2 million hectares) was also planned. Other initiatives were to include transport infrastructure, well-construction, animal health, work on crop storage and reforestation.

Franke and Chasin (1980) criticize this plan on a number of grounds, including its excessive focus on 'cash' rather than 'staple' crops (and a failure to appreciate the fluidity of such categories), and the excessive speed with which development is proposed, particularly in the field of irrigation. They argue that the plan too readily accepts simplistic assumptions about the nature and causes of overgrazing ('traditional' livestock raising practices), and fails to recognize the potential impacts of initiatives such as well-digging. Derman (1984) analyses the USAID component to Sahelian initiatives, and concludes that despite the New Directions mandate, projects showed little change. The 'familiar arsenal of modernization' was turned on the problem, the focus on small farmers being based primarily on the need to get the commercial structure right. He sees some suspicion of large schemes, and some interest in participation, but no effective appreciation of the links between ecology and the way production is organized. Franke and Chasin (1980) are more specific. They conclude that the plan would increase environmental damage, technological dependence, inequality and famine vulnerability. Their critical analysis of multinational investment in the Sahel and aid projects leads them to suggest that the vaunted 'new' approach to development in the Sahel would simply reproduce existing vulnerability to drought. Development, in this instance at least, would bring no solution to environmental degradation.

6

The Environmental Impacts of Development

Development and environmental pollution

The tight links that exist between poverty, development and environmental degradation are not an isolated feature of desertification policy. The creation of degraded environments cannot be seen as simply an unfortunate by-product of the development process. It is an inherent part of that process itself and the way in which development projects are planned and executed. Poverty and environmental degradation, driven by the development process, interact to form a perilous and unrelenting world for the poor. The environmental and socio-economic impacts of development initiatives are not always easy to separate. Nonetheless, this chapter focuses on the environmental side of impacts associated with development, looking at industrialial pollution, pesticides, and the physical and ecological impacts of river control. The next chapter discussses the ways in which development planning attempts to deal with these impacts, through attempts to reform the project planning process. Chapter 8 moves on from this reformist approach to argue for a more holistic and and radical view of environment. It draws out the links between environmental impact and those people affected, arguing that a people-orientated approach is necessary both to an adequate understanding of environmental impact and to attempts to identify a 'green' or 'environmentally sustainable' route to development.

Many development projects represent a response to a specific environmental problem, for example flooding or seasonal water shortage. Other projects are developed to tackle particular problems of poverty, or to take advantage of perceived economic opportunity. However development initiatives themselves are liable to have significant environmental impacts, and these will in turn impinge on social and economic conditions.

The role of development in creating degraded environments is probably most clearly seen in the context of pollution. This was one of the concerns which gave rise to the UN Conference on the Human Environment in Stockholm in 1972, but although at the time it was seen to be a problem of affluence and an industrialized economy, it is clear that it is also very much a problem of the Third World. The Founex Meeeting which preceded the Stockholm Conference (Chapter 3) adopted a broad definition of 'environment' (one subsequently adopted by UNEP). This recognized that it was the immediate need to tackle the problems of malnutrition, disease, infant mortality and illiteracy which set priorities for environmental action in Third World countries (Walter and Ugelow 1979).

Certainly Third World countries lack the environmental and anti-pollution safeguards that have become standard in the industrialized world (Park 1986). Walter and Ugelow (1979) present comparative data on environmental standards derived from a survey of 145 countries conducted in 1976 by the UN Conference on Trade and Development (UNCTAD). Only forty-two usable replies were received, seventeen from developed countries and twenty-five from developing countries. Environmental standards were scored from 'strict' to 'tolerant' in comparison with the USA as a benchmark. The developing countries had significantly more tolerant policies, although there was variation in both groups. Thus Singapore, Columbia, Chile and Israel, for example, had 'strict' environmental policies. There was an inverse relationship between level of development and rigour of environmental policies, but obviously there were many other factors: Walter and Ugelow (1979) point out, for example, that environmental quality is a matter of social choice, and hence societies may differ in priorities, and that environmental protection also reflects the nature of national political processes.

The most serious and intractable pollution problems in Third World countries arise in urban areas, both from the disposal of wastes and the by-products of industrial processes, for example in China (Zhao and Sun 1986), and Thailand (Tuntamiroon 1985). It is clear, for example, that the phenomenon of acid rain is serious in many parts of the Third World (McCormick 1985). However more mundane problems, such as sewage disposal (Lohani 1982, Ikporukpo 1983, Oladimeji and Wade 1984), and pollution from motor vehicles, are in practice just as intractable.

Problems of urban pollution exist on a particularly grand scale and acute form in in those countries in the Third World which have most successfully industrialized such as India, Malaysia, and Brazil. Cities such as Sao Paulo in Brazil are probably among the most heavily polluted environments in the world. It is ironic, of course, that it is these 'Newly Industrialized Countries', as the development commentators call them, which have been most successful in achieving the 'development' for which others strive. Pollution seems to have been part of the price paid for 'success' in development, or at least part of the price of the right to enter the development race. It seems to be an inescapable part of the development process. The economic costs of pollution are difficult to calculate, although perfectly real. Disasters such as that at Bhopal in 1984 reveal something at least of the human costs of industrial pollution.

Industrial pollution in the Third World is not simply an inevitable result of indigenous industrial development. Pollution can occur for reasons which relate to the poverty of experience and expertise, for example because the technology is lacking to clean up discharges, or the inappropriateness of certain Western technologies. Both explanations were forwarded when poison gas leaked from the pesticide plant at Bhopal, and both have some validity (Vaidyanathan 1985). So too does the view that inefficiency and corruption are inherent features of the planning and management of development projects in the Third World. Fundamental to that disaster, however, as to many lesser incidents of chronic pollution is the economic context of pollution.

The costs of pollution control tend to be passed either forwards to product prices (thus making manufactured goods more expensive), or backwards, reducing returns on capital. Pollution-control is therefore unattractive. It can therefore be argued that Third World countries without strict pollution legislation can achieve a competitive advantage over industrialized countries (Walter and Ugelow 1979). Lack of pollution controls (like cheap labour) cuts the cost of production, or rather transfers costs to the host environment and community. In a sense therefore, pollution is a hidden subsidy for Third World industry. The hazards and costs of industrialization without pollution control can therefore seem to have a certain attractiveness. When this is combined with the inertia, inefficiency and corruption with which Third World government bureaucracies are plagued, the result can be - as in the case of Bhopal - disastrous.

Low pollution control standards in the Third World are also greatly advantageous to First World companies faced with rising production costs at home. Transnational companies in particular have moved industrial plants that are highly polluting to locations in the Third World as a direct result of environmental protection policies in industrialized countries. Thus Suckcharoen *et al.* (1978) discuss the location of a caustic soda factory in Thailand by a Japanese company in 1966 because of weak local environmental protection legislation. Their concern is the danger of pollution by methyl mercury in aquatic ecosystems, and especially in fish, which form 50 per cent of the animal protein intake in the average Thai diet. Mercury poisoning among the fishing community at Minimata in Japan (causing extensive birth defects in children) was a major factor in raising consciousness of the dangers of industrial pollution in industrialized countries (D'Itri and D'Itri 1977). Suckcharoen *et al.* (1978) show that mercury levels in fish near the caustic soda factory are on average 1.48 ppm compared to low background levels of 0.07 ppm. Levels in human hair are still low, but they predict serious future problems.

Similar examples of pollution from plants built or run by First World based multinationals are legion. Walter and Ugelow (1979) present data on the percentage of capital expenditure spent on pollution control by US-based multinationals in the US and overseas. In primary metals industries pollution control comprised 21.1 per cent of domestic capital outlays, but only 9.8 per cent of overseas capital, and proportions were similar for other industries, for example chemicals (8.9 per cent in USA, 5.3 per cent abroad) and pulp and paper (21.9 per cent in USA, 11.8 per cent abroad). Castleman (1981), for example, describes piles of asbestos-cement waste and the discharge of untreated waste water outside a factory in Ahmedabad in India making asbestos-cement building materials.

Often it is the exploitation of primary resources which creates the most extensive problems. Certain primary industries are highly polluting, notably perhaps the manufacture of pulp and paper. Christiansson and Ashuvud (1985) discuss the potential environmental effects of a paper mill in Mufundi District in Tanzania. Petr (1979) describes the prospective impact of a copper mine on the Ok Tedi River in Papua New Guinea, which included the downstream transport of mine tailings and associated heavy metals. He remained sanguine about the possibility of dealing with the resulting impacts, but elsewhere serious problems have developed following development of this kind. Baluyut (1985) describes the impacts of three gold and copper mines on the Agno River in the Philippines. These use cyanidation and flotation to obtain ores, and produce large volumes of tailings which are impounded in tailing ponds. Failure of the retaining dams of these ponds has resulted in the release of tailings comprising fine sediments and toxic materials into the Agno River. Downstream ecosystems, particularly coastal mangroves, have been adversely affected by sedimentation, and heavy metals (mercury, cadmium and lead) have accumulated in fish and shellfish downstream.

The treatment of bauxite with caustic soda to produce alumina creates similar toxic waste disposal problems, notably perhaps in Jamaica, where 13 million tonnes of caustic alkaline slurry are produced every year by the transnational-dominated aluminium industry (Bell 1986). Ikporukpo (1983) describes the problem of crude oil pollution in the Niger Delta of Nigeria; its impacts include loss of crops, sterilization of soils and the death of fish. In the decade 1970-80 Nigeria experienced eighteen major oil spillages involving over 1 million barrels of oil. The economic costs, and the disruption to production and economy in the Niger delta, were considerable. For example, in 1970 0.8 million Naira (£30.6 million) were paid in compensation for a single incident in Rivers State.

The environmental impact of pesticides

Not all pollution problems in the Third World are associated directly with industrialization. One area of growing concern is the use of pesticides, both in agriculture and the control of disease vectors. Crop losses through pest attack are a major problem in the Third World, both during crop growth and during storage. It is estimated, for example, that in-field crop losses are 41.6 per cent in Africa, 13 per cent from insects, 12.9 per cent from disease and 15.7 per cent from weeds (Ghatak and Turner 1978). Losses in Asia are 43.3 per cent, in Latin America 33 per cent. The attractiveness of pesticides (the word embracing insecticides, herbicides, fungicides etc.) to those determined to increase food and commodity production is obvious. The resurgence of locusts in Africa in the mid-1980s following the temporary reduction in chemical control measures and the disbanding of the Office International Contre la Criquet Migrateur Africain in 1986 has proved graphic and emotive evidence for many observers of the need for continued chemical control of crop pests (Jago 1987).

The distribution and sale of pesticides is almost entirely in the hands of transnational agribusiness companies, who eagerly promote the advantages of chemical pest control, yet there are questions about the economic viability of pesticide use in Third World Agriculture. It is relatively easy to measure the direct costs of pesticide use, but much harder to estimate the benefits (and hence the costs of damage avoided by particularly prophylactic use before pest outbreaks occur). It is also difficult to obtain data on pollution effects. These include problems of the impact of pesticides on non-target organisms and the development of resistance of target organisms.

At a national scale Ghatak and Turner (1978) stress the need to consider the foreign exchange costs of importing pesticides (for few countries in the Third World have indigenous production capacity), and point out that pesticides often serve as substitutes for labour inputs (for example in weeding), and labour is neither expensive nor in short supply in much of the Third World. Certainly the substitution of pesticides purchased for cash for family labour is unlikely to make sense in most peasant households. Cox (1985) studied the economics of pesticide adoption and use on a variety of crops in Tanzania. Both the widely-used pesticides, DDT and inorganic copper fungicides, are cheap. They are both persistent in the environment, and Cox points out that the implications of this - for example on the use of drainage water for drinking downstream - are not known. Long-term social costs cannot therefore be calculated. He discusses the problems of using conventional cost-benefit analysis, and suggests that pesticides should only be used when private benefits (that is to the user) exceed private costs and all immediate social costs.

This is a pragmatic approach, but the fact remains that the longer-term environmental and social costs are important. The problems of pest resistance to pesticides under continuous use is well-documented (Bull 1982). The most commonly quoted example of pesticide resistance is not in agriculture, but in public health: the attempt by the World Health Organization (WHO) to eradicate Malaria using DDT (an organochlorine compound) and drug treatments on sufferers in the 1950s. Initially this met with enormous success, the number of cases being reduced from an estimated 300 million pre-1946 to 120 million in the late 1960s, a period when world population had greatly increased. However, from the mid-1960s mosquitoes resistant to organochlorine pesticides began to appear, and the number of new cases of malaria rose once more (Bull 1982). Resistance to the prophylactic drugs taken against the malaria parasites are also now a major problem in some areas of the Third World, particularly those areas popular with tourists such as the Kenyan coast.

There is also evidence that the use of organophosphorus and carbamate insecticides in agriculture is implicated in some instances in the development of pesticide resistance in mosquito vectors of malaria (PEEM 1987). Resistance can appear quickly in a target population. The Onchocerciasis Control programme in West Africa began spraying the organophosphorus insecticide temephos to control *Simulium damnosum*, the vector for river blindness, in 1974. Since 1974 the project has sprayed 18,000 km of river per week in the wet season and 7500 km per week in the dry season in Upper Volta, Mali, Ivory Coast and Ghana (Walsh 1985). Resistance was first recorded on the lower Bandama river in the Ivory Coast in March 1980, but spread rapidly to surrounding areas. The project switched to chlophoxim (another organophosphorus compound), but resistance to this appeared by mid-1981 and this too was withdrawn. It appears that resistance to one organophosphorus pesticide promoted resistance to others (Walsh 1985).

In Malaysia the number of rice pests resistant to at least one pesticide rose from eight in 1965 to fourteen in 1975, the brown plant hopper (*Niliparvata ingens*), for example, bouncing back in the late 1970s in the face of growing and intensifying pesticide use (Bull 1982). It is now clear that the brown plant hopper is controlled by a wide guild of predators, and rapid resurgence follows accidental control by pesticides of these organisms. There is now an FAO/UNEP

Cooperative Programme on Integrated Pest Control in South East Asia tackling this problem, trying to reduced levels of pesticide use and devise integrated approaches to pest control including breeding rice varieties with pest-resistant attributes (PEEM 1987). Similar problems emerged with cotton spraying against whitefly on the Gezira Scheme in the Sudan. The area sprayed increased rapidly from 600 hectares in 1946 (0.7 per cent of the total area) to 98,000 hectares (100 per cent) in 1954. The number of sprays then increased from one per year in the late 1950s to nine per year in the late 1970s. By 1976, 2500 tonnes of insecticide were being used per year at Gezira, yet despite this effort, and the massive rise in production costs it represents, pest losses continued (Bull 1982).

Although studies of the effects of pesticides on crop yields are numerous, there are few studies of impacts on other aspects of the ecosystem. Perfect (1980) analysed the effects of DDT on fields in Nigeria. He found DDT accumulated in the soil over the four-year experimental period, although at lower levels than those in temperate areas (2 per cent of that applied) because of more rapid volatilization. Soil organisms showed shifts in numbers and species complement, and there were both qualitative and quantitative changes in the pathways and rates of litter breakdown. Although yields were higher in the short-term under pesticide use, contamination in soils following pesticide use reduced yields. He suggests that DDT may impair the system's capacity to regenerate fertility in fallow periods.

Important though the problems such as pesticide resistance are, the strongest focus of criticism of the use of agro-chemicals in the Third World has been the problem of accidental poisoning. Third World countries account for only 15 per cent of global pesticide use, but their consumption is rising rapidly, much of it sponsored by First World development aid. Over half the cases of pesticide poisoning occur in the Third World (Bull 1982). It is estimated that there are 375,000 cases of poisoning by pesticides in the Third World each year, 10,000 of them fatal (Caufield 1984). In Vavuniya in northeast Sri Lanka, for example, there were 938 deaths from pesticide poisoning in 1977, more than malaria, tetanus, diptheria, whooping cough and polio put together (Bull 1982, p. 44).

The causes of this toll in the Third World are fairly clear: product labelling is poor and inappropriate, often not in a relevant language, and often without any provision for use by illiterate users, who are of course the majority. Farmers are unable to read warnings about toxicity, and lack the knowledge to interpret dangers, or the equipment to apply pesticides safely. It is quite unrealistic to expect rules of application devised in developed countries, for example using clothing proof against sprays and protective masks, or washing before eating, to be practicable in the rural Third World. Medical support in the case of poisoning is rarely obtainable.

Studies in Central Luzon, North of Manila in the Philippines, showed major rises in pesticide purchases in the early 1970s, accompanied by a 27 per cent rise in the death rates of men of working age Pearce (1987). Deaths up until 1976 peaked in August, the height of the spraying season. In that year double-cropping began on the local irrigation scheme, and spraying in February was matched by a second peak in deaths in that month. Protective clothing was not worn, and the backpack sprayers used put 40 mg of active ingredients of pesticide onto operators per hour (Pearce 1987).

Pesticides also represent a serious problem of hazardous waste disposal, or when used wholly outside their intended context. In Guyana, the rodenticide thallium sulphate was imported in 1981 to kill rats in sugar cane plantations. It was misused as a fertilizer and in other ways, and contaminated milk and grain caused a number of deaths (MacKenzie 1987). Pesticide use rose in the Pacific, as elsewhere, in the 1970s, and there have been a number of reported cases of the death of fish from pesticide spills, for example of lindane and DDT in the lagoon of Tokelau, or the leakage of endrin and sodium arsenate into streams feeding the lagoon on Yap (Brodie and Morrison 1984).

There has been considerable debate about the use of pesticides in Africa to control tsetse flies (*Glossina*), which carry sleeping sickness, and an equivalent disease of livestock, nagana. Over 10 million square kilometres of Africa, in thirty-four countries, have tsetse. Attempts at eradication began early in the twentieth century through campaigns to exterminate game which harboured the trypanosome parasite, followed up in the 1920s and 1930s with the clearance of bush habitat (Matthiessen and Douthwaite 1985). In Uganda over 17,000 square kilometres of bush was cleared, in Tanzania over 21,000 square kilometres. In the 1940s aerial dusting with organochlorine pesticides DDT and gamma-BHC (lindane) began in central Africa. This was replaced with spraying of these pesticides and dieldrin, both indiscriminately and from the ground using manual or vehicle sprayers on tsetse resting sites. Over 300,000 square kilometres have been sprayed with organochlorine compounds. More recent methods include fine aerosol spraying from aircraft of large areas, and the introduction of synthetic pyrethroids (Matthiessen and Douthwaite 1985), although there are a series of new approaches using traps coated with pesticides baited with chemicals attractive to tsetse (Alsopp *et al.* 1985), and sterilization and release methods (Redfern 1986). Despite past failures, the intention of the current FAO-coordinated project is to eradicate the tsetse from Africa completely.

The environmental impacts of tsetse control are almost as complex as the evolution of control policy, and the science which drove it (Ford

1971). There is still only a limited amount of information on the ecological impact of continued broadcast pesticide use on savanna ecosystems. Recent comment by environmentalists has focused on possible problems, on the implications of tsetse removal on land use, and particularly the problems of land degradation which might arise from intensified cattle management (Linear 1981, Matthiessen and Douthwaite 1985, Ormerod 1986, Kemf 1987). The other major aerial spraying programme in Africa, the Onchocerciasis Control Project, has been described above. It has attracted less attention from environmentalists than the tsetse programme, although it is also on a remarkable scale. Organophosphorus pesticides temephos and chlorphoxim were used to minimize problems of DDT of impacts on non-target organisms and persistence in the environment. It is argued that environmental impacts were acceptable (Walsh 1985).

The danger of poisoning of those applying pesticide, whether in agriculture or disease control, has become more acute with the transition from organochlorine pesticides to organophosphorus compounds. The organochlorines, such as DDT and dieldrin, were the subject of much controversy in Britain and other industrialized countries in the 1950s and 1960s because they do not metabolize, but become stored in fatty tissue. There were a number of direct poisoning incidents on farmland birds in the 1950s, and subsequent research showed that in top predators such as birds of prey concentrations were reached which caused physiological problems such as eggshell thinning. There were serious population declines in some species (Moore 1987, Sheail 1985). The big advantage of organochlorine insecticides however, apart from their cheapness, is that they are relatively safe to human users. Their replacements, organophosphorus pesticides such as pirimphos-ethyl, endosulfan or disulfoton, have a high dermal toxicity, that is they are poisonous when in contact with the skin. Disulfoton for example inhibits the production of cholinesterase, an enzyme important in nerve function, lack of which causes convulsions. The choice of pesticide to minimize risks to environment and people is therefore far from easy.

As controls on pesticide use tighten in the industrialized world, Third World markets are increasingly attractive to producers, particularly when pesticides banned at home can be exported to countries where environmental controls are more lax. In 1979 25 per cent of the pesticides exported by US companies were either banned or unregistered in the USA (Caufield 1984). Britain is the world's third largest exporter of pesticides, supplying 15 per cent of pesticides in international trade, and the industry faces a similar challenge to make its sales responsible and humane. The number of chemical compounds on the market is rising rapidly, by up to 100 per year, and there is a clear need for international cooperation to list those products banned in different countries. The UN General Assembly finally instructed the Secretary General to compile such a list in 1982 (Shaikh and Nichols 1984).

The expansion of pesticide use in the Third World is an integral part of the development process. It is sanctioned and promoted by development agencies, and financed by First World loans. The industry is run from the industrialized world, and the expansion of pesticide use in the name of development is good business. There are both ecological and economic costs to pesticide use, although these are difficult to identify, and it is clear that at certain times and in certain places these costs will outweigh the benefits of increased yields and disease control. There are also human costs, particularly from those who apply poisons unprotected and in ignorance of their toxicity, and these are not always borne by those who stand to benefit. Higher productivity may mean cheaper food for urban consumers, but it may also mean lower income and perhaps an early death for a peasant farmer. Pesticides may raise yields, but if they also raise the cost of production, and take the farmer onto a treadmill where pest resistance demands new and larger pesticide applications, their benefits may be a cruel illusion. Pesticides, like other aspects of development, are not magic. Like industrial pollution, they are just one element in the complex equation linking poverty and environmental quality.

Development and degradation in rainforest environments

The most emotive demonstration of the links between development and environmental degradation is undoubtedly the question of the development of rainforest, or more properly and less confusingly tropical moist forests (Myers 1980). The tropical moist forests are the subject of a stark and insistent picture of destruction ably and emotively painted by environmentalists (e.g. Myers 1984, Caufield 1985) and many scientists. In 1973 Denevan suggested that development was bringing about 'the imminent demise of the Amazonian rainforest' as a result of a 'pell-mell destructive rush to the heart of Amazonia' (Denevan 1973, p. 137). His alarming analysis was criticized, but he argued in 1980 that further data confirmed its conclusions (Denevan 1980).

There is certainly no lack of shocking statistics, both about Amazonia and tropical moist forests elsewhere. The Friends of the Earth Tropical Rainforest Campaign which began in 1985 presented data that 7.5 million hectares of undisturbed tropical moist forest were destroyed or degraded annually, making 30 acres cleared every minute, and suggesting that by 1990 the rate of extinction of species will have risen globally to one per hour (Secrett 1985a). Data are, however, surprisingly limited, and difficult to interpret. This is partly because of difficulties of definition, both of the forest types and what is meant by 'deforestation'.

There is by no means universal agreement on the extent of tropical moist forests. The FAO recognizes closed forest (1.20 billion hectares globally) and open forest (0.73 billion hectares). However, they also include fallows of both closed forest (0.24 billion hectares), and open forest (0.17 billion hectares), and scrubland (0.62 billion hectares), (Lanly 1982). There are significant differences between the forests of different continents. Over 56 per cent of closed forests are in Latin America, with 45 per cent of fallow closed forests, while Africa has 66 per cent of open forests, 61 per cent of fallow open forests and 71 per cent of scrub.

Confusion over the extent of forests is greatly compounded by differing estimates of the extent of deforestation (Allen and Barnes 1985). One cause of the discrepancies which exist is lack of a clear and universal definition of deforestation. Variations in terminology lead to what appear to be substantial differences in estimates of rates of forest disturbance, although it is argued that these can be reconciled (Melillo *et al.* 1985). Certainly, where large tracts of land are cleared permanently of forest and turned over to pasture, definition is easy. However it is far more difficult, for example, to decide when to count internal changes caused by shifting cultivation practices as forest clearance.

One possible response to the lack of data on deforestation is the use of satellite imagery (Green 1983, Myers 1980). Experiments suggest that such data have a potential usefulness. Singh (1986) used data from the Landsat Multi-Spectral Scanner (MSS) to assess the extent of clearance of forest in Manipur State in India. MSS data has a higher spatial resolution but was only available at one point in time. It was of limited use, partly because of cloud cover, and partly because the clearance taking place was primarily by shifting cultivators, and involved small areas of complex shape. Cleared land remained vegetated, and the relatively small changes in spectral response were not readily discerned. Landsat data are also used by Hamilton (1984) in his study of deforestation in Uganda, although as part of a much larger historical study which also makes use of other sources.

Tucker *et al.* (1984) use data from the Advanced Very High Resolution Radiometer, which has a coarse spatial resolution but is available on a daily basis, to reveal strip clearance of forest in Rondonia in Brazil for roadside settlement. Fearnside (1986) analyses Landsat data to derive estimates of the extent of forest clearance in the Brazilian Amazon, and argues that although the total extent cleared is still small compared to the 5 million hectares of Brazil's Amazon region, the rate of clearance accelerated rapidly between 1975 and 1980, and expanded exponentially in certain states. He concludes that deforestation is rapid, and that arguments to the contrary are wrong. Clearance is also highly concentrated. In certain key areas, notably along the Belem-Brasilia Highway through southern Pará and northern Goiás, and along the Cuiabá-Porto Velho Highway through Mato Grosso and Rondonia forest clearance is intense.

Allen and Barnes analyse the various global figures of change in forest area over time. Although there are significant differences, the overall trend in forest extent is firmly downwards (Allen and Barnes 1985). Data for twenty-eight African, Asian and Latin American countries show that rates of deforestation between 1968 and 1978 are correlated with population growth, fuelwood production and wood export, and agricultural expansion. They are not correlated with growth in GNP per capita, and hence with 'development'. As work such as that of Hamilton (1984) shows, global generalizations of this sort are of limited value. What is needed is a detailed national and regional survey. The simple global picture is complicated by the existence of marked regional variations in both the nature and extent of forest. These are reflected in turn in the diversity of rates of deforestation, and the factors causing them (Myers 1980).

The main causes of loss of tropical moist forests are forest farmers, the timber trade, cattle raising, and (to a much lesser extent) fuelwood. Farmers are probably the most significant anthropogenic influence on tropical moist forests, particularly in Africa and Latin America. In the mid 1970s forest farmers numbered perhaps 140 million, and occupied over 20 per cent of the area of tropical moist forests (Myers 1980).

Timber production (largely for European and South East Asian markets) is the dominant factor in deforestation in South East Asia, particularly Malaysia, Indonesia, Papua New Guinea and the Philippines. In comparison with this, Latin American forests seem relatively unimportant in terms of international trade, but it should be noted that there is a considerable internal trade in timber in Brazil.

The environmental impacts of deforestation depend to a considerable extent on subsequent land use, which can range from abandonment following timber clearfelling through replacement with a pasture for cattle raising, establishment of a plantation of tree crops, to an agro-forestry system. There is now a considerable literature on impacts, focusing on both local and long-distance or (in the case of the link drawn between rainforest loss and the rise of atmospheric carbon dioxide) global ecosystems (e.g. Sutlive *et al.* 1980, 1981, Sioli 1985). The most obvious immediate impacts are the interruption of the rapid nutrient cycling of tropical moist forest ecosystems and the resulting loss of nutrients caused by the low retention capacity of forest soils (Herrera *et al.* 1981), the exposure of soil to erosion, and the effects of faster and more sediment-laden runoff downstream (Lal 1981). These impacts are most clearly seen in the case of the Amazonian rainforest cleared for cattle ranches and cultivation, although directly similar effects occur with clearfelling logging for timber.

The expansion of settlers and ranchers with the support and encouragement of the state into the Amazon has been relatively recent, but it has attracted a great deal of critical comment (e.g. Goodland and Irwin 1975, Goodland 1980a, Eden 1978, Moran 1983, Hemming 1985, Caufield 1985). Allen (1983) argues that the supposed 'virgin lands' of the Amazon have acquired an image which owes little or nothing to the substantive success (or more usually lack of it) shown by any actual development projects. They are sustained by fantasies woven about them and sustained by government rhetoric. Skillings (1984) also cites the psychological effects of the sheer extent of Amazonia; this forms 57 per cent of Brazil yet holds only 4 per cent of her population and generates 2 per cent of GDP. He suggests this acts as a challenge to political leaders, one backed up by perceived strategic implications.

Others offer more prosaic explanations, for example Hecht (1984, 1985). She rejects simple models of degradation because of population growth (a neo-Malthusian model), the requirements of export crops, and the inappropriateness of agricultural technology (the 'false paradigm' model), and argues instead for an understanding based on the political economy of Brazil. Following the military coup in 1964, legitimacy was sought through economic output. The astonishing growth in the Brazilian economy in the 1960s and 1970s was accompanied in the rural sector by increasingly industrialized methods, and a crisis of access to land by the poor. Hecht argues that Amazonian development obviated the need for land reform. Furthermore, new technologies of seeded pastures opened up profitable possibilities for industrial agribusiness to move into beef ranching, thus providing an important (and largely new) export product, new forms of production, and an outlet for investment. Hecht argues that the needs and incentives within Brazil which pointed to the development of Amazonia were supported by international concerns and interests, for example the views of the World Bank and the FAO.

In 1965 Operation Amazonia was launched. Laws passed in 1966 provided tax reductions for companies (including foreign corporations) investing in Amazonia, and additional credit was available both from within Brazil and internationally. Ranching was the dominant activity, and land prices began to rise. In 1970 the New Integration Programme (PIN) brought a new focus on small-scale settlement, but the policy was again reversed in 1975 and ranching predominated once more. Cattle ranching has achieved notoriety through critical accounts of 'the hamburger connection' (Myers 1981), and the linkage between

Amazonian forest clearance and American fast food outlets. American beef prices rose in the 1970s, and by 1978 Central American beef was less than half the wholesale price of beef raised in the USA. American beef imports rose rapidly, the lean grass-fed beef of rainforest pastures being suited to the fast-food industry, which comprises 25 per cent of US beef consumption. Between 1961 and 1978, beef exports rose by five times in Costa Rica and by fifteen times in Guatemala (Myers 1981). In Brazil the dry *terra firme* land became the subject of intensive development both by transnational companies such as Volkswagen, King Ranch and Armour-Swift, and also large investors from Brazil (Fearnside 1980).

There is a pervasive belief among both scientists and policy-makers in Brazil that cattle ranching is a valid and effective long-term land use, and indeed that the creation of pasture improves soils (Fearnside 1980). This is based on the argument that clearance of forest releases nutrients held in rainforest vegetation, and there is a rapid increase in nutrient availability. In fact, critics of cattle ranching in rainforest-derived pastures argue that this view ignores serious longer-term ecological impacts, which make cattle ranching far from sustainable (Fearnside 1980, Hecht 1980, 1984, 1985). Forest clearance does generally bring about short-term falls in soil acidity, and a rise in the amount of calcium, magnesium, potassium and phosphorus, although there is a great deal of variability and larger samples tend to show that increases are less dramatic than is often claimed. However, the pulse of nutrients is short-lived. In particular, phosphorus levels decline after about five years to something very close to the levels in soils under undisturbed forest, and over longer-term five- to fifteen-year periods nutrient status and pasture productivity decline, sometimes drastically. Soil erosion can also become a serious problem.

Cattle ranching is therefore not usually a sustainable land use in rainforest areas (Fearnside 1980). Ranch economics dictate the sale of pasture after about five years, indeed Hecht (1985) reports that 85 per cent of ranches in Paragominas had failed by 1978. In ecological terms this seems to make little sense, but of course such short-term exploitation of land can be good economics. Entrepreneurs are attracted to rainforest ranching not by the long-term productivity of the land but by short-term returns on investment (Hecht 1984, 1985). As she comments, 'if the productivity of the land itself is of low importance, cautious land management becomes irrelevant and environmental degradation is the inevitable result' (Hecht 1984, p. 393).

Allied with this environmental degradation are impacts on smaller producers. Bunker (1980) describes the impact of large-scale ranching on small peasant producers on *terra firme* in Médio Amazonas. Peasant producers often lacked title to land, and were unable to sustain production in the face of competing demands from ranchers for accessible land as land prices rose. Ranching therefore appears to have been socially regressive as well as environmentally damaging. However, small shifting cultivators and small-scale ranchers are themselves major agents of destruction of rainforest, both in Amazonia and elsewhere. Indeed, globally they are believed to be a greater threat. Myers (1980) suggests that 20 million hectares of tropical moist forest land is damaged or destroyed for agricultural land by or for farmers. This is an order of magnitude more than that cleared for ranching, although the spatial concentration of ranching in the Latin American forests makes that a particularly acute phenomenon. In East Kalimantan (Indonesia), shifting cultivation has created 2.4 million hectares of secondary forest and 0.4 million hectares of grassland dominated by *Imperata cylindrica* grassland (Kartawinata *et al.* 1980). This amounts to 16 per cent of the total forest area. Forestry staff lament illegal cutting by people 'who do not understand the ecological consequences of forest destruction' (Soeriangara 1982, p. 82).

The impacts of forest clearance for agriculture are complex (Nye and Greenland 1960), and relate chiefly to the land use with which the forest is replaced. Hecht (1980) reviews impacts in Amazonia, and compares them with those of forestry and ranching, Kartawinata *et al* . (1980) describe succession following slash and burn in Indonesia. To simplify greatly, there is once again a flush of nutrients after clearance, followed by a fairly rapid decline in soil nutrient status and yields. It is clear that the use of annual crops such as maize are not sustainable, and indeed most rainforest farming depends on shifting cultivation. Where access to land is restricted, population pressure rises, or farmers are trapped by institutional constraints such as indebtedness to a settlement scheme authority, inappropriate agricultural techniques lead rapidly to exhaustion of soil nutrients (Trenbath 1984), low economic returns and depauperation of farming households. The economic and ecological collapse of such systems is not unusual. It is not surprising that settlement schemes in rainforest areas have proved difficult to plan, and are of limited success even in narrowly economic terms (Schmick and Wood 1984). Hiraoka and Yamamoto (1980), for example, are critical of the way settlements in northeast Ecuador are ill-suited to the environment. Many projects, such as the enormous Transmigrasi Project of Indonesia, to resettle people from Java (Hardonjo 1977, Otten 1986), appear to derive their justification primarily from powerful political support (Budiarjo 1986).

Some indigenous shifting cultivation methods are sustainable, at least at low levels of intensity. Perceptions of the need to convert indigenous forest farmers to 'modern' methods of agriculture based on continuous cropping are the origins of sometimes draconian development policy. The extent of shifting cultivation among tribal people in forests in northeast India, and the feasibility of making such a transition, are discussed by Thangam (1982). In rainforest environments various attempts are now being made to define cash-crop regimes aimed particularly at settlers which either mimic the physical structure of rainforest or include agroforestry elements to enrich soil nitrogen and phosphorus, and hence allow sustainable continuous cropping (Nicolaides *et al.* 1984). There are obviously economic as well as ecological constraints on developments such as this, for example low world demand for possible tree crops such as cocoa (Goodland 1980b).

The other main cause of deforestation is logging, and again environmental impacts vary considerably with the logging regime adopted. Traditional selection logging, of the kind still practised in the late 1970s along the Urumbamba River in eastern Peru (White 1978) have died out in most areas of tropical moist forest. In East Kalimantan (Indonesia), for example, 13 million hectares out of a total of 17.3 million hectares of dipterocarp forest are allocated for mechanized logging (Kartawinata *et al.* 1981). Logging of natural forest can broadly be classified as selective or clearfelling, but within this there are distinctions. Selective logging intensities can vary from two to three stems per hectare taken out in some African forests to twenty or more stems per hectare in parts of South East Asia. This difference is caused less by the attitude of foresters to conservation than by the density of timber of suitable size and quality.

Both selective felling and clearfelling have extensive impacts on forest ecology, although of the two clearfelling is obviously the more destructive. The two may be operated together in different zones of the same forest, and worked forests may exhibit a complex and fragmented structure with some areas logged at near 100 per cent and others with a canopy reduced by 30-50 per cent. Selective logging can take place on a monocyclic basis (a single operation to remove all saleable trees), or in a polycyclic system where trees are removed in a series of felling cycles as they reach suitable sizes (Johns 1985).

The nature of the logging process obviously affects environmental impact, as does the nature of post-logging treatment. Typical problems, however, include soil compaction and loss of internal structure, effects of leaching and runoff on soil nutrients and micro-organisms, the loss of nutrients in timber, and changes to the internal microclimate of the forest (Shelton 1985). Ground vegetation is lost through desiccation, and animal and bird species dependent on shaded forest conditions are also lost. Wildlife species such as primates are unable to survive in heavily logged forests, although a limited number of species can survive selective logging (Johns 1985). Territorial species such as gibbons are tenacious in surviving forest fragments (Shelton 1985). Studies of the ecology of forest fragments are being made in Amazonia to assess exactly what survives in such refugia left after widespread deforestation (Lovejoy *et al.* 1983).

In East Kalimantan, intensive selective felling of fourteen stems per hectare by mechanized methods caused damage to 41 per cent of residual trees (Kartawinata *et al.* 1981). Skid tracks, haul roads and log yards occupied 30 per cent of the logged area. The recovery of these compacted and scraped areas is slow, water infiltration rates are low and erosion is high. Selective logging also involves the loss of the best specimens of commercial species, and sometimes the extinction of species, forms of 'genetic erosion' which have significant impacts on the physical and floristic structure of the forest (Kartawinata *et al.* 1981).

One area that has been intensively studied is the Gogol Valley, Madang province, Papua New Guinea. Felling for woodchips and saw logs began in 1973 on a logging concession of 68,000 hectares in the Madang and Naru Valleys, held by a local operating company of the Japanese company Honshu Paper, part of the Mitsui group (Lamb 1980, Seddon 1984). By 1983, 36,750 hectares had been logged. Little of the logged land was cultivated. The resulting forest regeneration was studied by Saulei (1984). Logging caused the removal of humus and topsoil, leading to high rates of soil erosion and leaching, deposition of sediments and waterlogging on valley floors and flats with associated soil acidity, and the loss of phosphorus. There were significant effects on the floristics of the forest which eventually regenerated, partly because of the loss of soil seed sources. Cultivation or other disturbance (such as loading areas) prevented regeneration of trees, and grassland of *Imperata cylindrica* and *Saccharum* species took over. However, as Saulei points out, regeneration of secondary forest is taking place. Furthermore, he argues that the forest cleared was itself a secondary growth resulting from drought and fires in the 1930s and during 1944-5. The considerable environmental effects may therefore not prove as disastrous as some fear.

The Gogol Valley development in Papua new Guinea is interesting, not least because from 1975 to 1978 it was made the subject of Man and the Biosphere funded research. Workshops were held at Port Moresby in 1975 and Madang in 1978, and research helped highlight problems in the area. However, the initiative was not coordinated, and no further formal work was done after this (Seddon 1984). The Gogol project was seen in some quarters as a showcase of forestry management (Lamb 1980). Initially the project was intended to integrate sawn timber, veneers and woodchips for pulpwood. The plan was for 48 per cent of the land to be clearfelled, 22 per cent selectively felled and 30 per cent unlogged (Lamb 1980), and the project was to include an element of reforestation. In practice, reforestation fell far short of the 800 hectares per year target, only 1000 hectares being planted in the first five years, mostly with species of *Eucalyptus*, *Terminalia* and *Acacia*. The rate of planting subsequently increased, but the extent of plantation remains too little to sustain the pulp mill after the fifteen to twenty-five year life of the project, and in practice it has been heavily subsidized in the form of research expertise by the Papua New Guinean state. Timber volumes in the forest have proved 15 per cent less than expected. Without new leases, it was estimated that the chipping operation would close by 1990 (Seddon 1984).

The economic benefits and costs of the Gogol scheme are hard to calculate (Seddon 1984). The Gogol Valley is home to 2,000-4,000 indigenous people, who lived by shifting agriculture, and for whom forms of permanent agriculture have been sought. Royalties paid to the indigenous people have certainly injected cash into their economy. Experience of other promised benefits (access, education, jobs) has been mixed. For example, logging roads have not survived the extraction operation. Against this must be balanced the trauma of the loss of traditional forests (although areas round villages and river corridors were spared), and the economic fact that the holding company JANT failed to show a profit, and thus escaped taxation. Profits were written off against capital borrowing costs and (it is alleged) transfer pricing allowed profits to be repatriated (Seddon 1984).

The physical impacts of river control

Rainforest development is by no means the only area in which the environmental impacts of development have been widely identified as a cause for concern. Another is the construction of dams on tropical rivers. A series of major dam projects were begun in the Third World in the 1960s. Most notable among these were perhaps those in Africa, the Aswan High Dam on the Nile, which was begun in 1960 and finished nine years later (Greener 1962, Little 1965, Waterbury 1979), the Kariba Dam on the Zambezi, Akosombo on the River Volta, and Kainji on the Niger (Hart 1980, Mabogunje 1973, Adams 1985c). The reservoirs created by these dams became the subject of considerable international research attention (Lowe-McConnell 1966, Obeng 1969, Rubin and Warren 1968, Ackerman *et al.* 1973). Furthermore, their ecological impacts became one of the foci of environmental concern at that time about development in the Third World, for example in the volume *The Careless Technology* (Farvar and Milton 1973). This concern grew rapidly and with an astonishing unison, as critics identified and explained failings in the way dam projects were conceived and designed (Sulton 1970). Tropical dams have never quite lost the notoriety among environmentalists acquired at that time (e.g. Linney and Harrison 1981, Goldsmith and Hildyard 1984).

The perception of that period was simple. Dam design failed to take into account secondary effects. Lagler (1969) described the problems of economic loss and human suffering which could arise where planning failed to look far enough ahead. Sulton described problems of 'myopia in the planning process, misinterpretation of ecological signs, poor timing and indifference to human suffering' (Sulton 1970, p. 128) which observers suggested accounted for many of the unfortunate yet often avoidable consequences of the construction of dams. The problem was one of planning failure, or in Hall's terms 'planning disaster' (Hall 1980). It was seen to be caused by the failure to integrate existing knowledge of ecosystems, specifically what Lagler (1971) referred to as concepts of total ecosystems, into the planning process. Such ecosystem thinking was still relatively new in the 1960s (McIntosh 1985). It proved very suited to the elucidation of the various complex relationships involved in river and floodplain environments, and there is now a substantial body of knowledge on the ecological effects of river control (e.g. Oglesby *et al.* 1972, Baxter 1977, Ward and Stanford 1979, Petts 1984).

The impacts of dam construction on downstream hydrology and erosion have been intensively investigated (e.g. Gregory and Park 1974, Buma and Day 1977, Park 1981, Petts 1984). Most research in these techical fields has taken place in the temperate rivers of developed nations. However, there are studies of other areas, for example by Chien (1983) on the effects of dam construction on the regime of the Han River in China, and Pickup (1980) who modelled hydrology and sediment yield of the proposed Wabo scheme on the Purari River in Papua New Guinea. Sagua (1978) identified a 60 per cent reduction in downstream flows between 1970 and 1976 following construction of the Kainji Dam on the River Niger, although the Sahel drought undoubtedly had an effect during this period.

The nature of hydrological effects varies with the purpose of the dam and the seasonal regime of the river. However, moderation and delay of the incoming flood peak frequently takes place because of the flood-routing effect of the storage impoundment. Such effects can be particularly significant where river regime is flashy and such peaks are common, for example in rivers in the semi-arid tropics. Flood-control dams exacerbate peak flow moderation effects, particularly in such seasonally torrential rivers. Dams for all-year irrigation moderate variations in flow regime on a longer timescale, storing water at seasons of high flow for use at times of low flow. However, unless they also have a flood control function, discharge beyond storage capacity is usually spilled, allowing some flood flows to pass downstream, albeit in a routed and hence attenuated form. Hydroelectric dams are designed to create a constant flow through turbines, and therefore tend to have a similar effect on discharge patterns. However, if the intention is to provide power at peak periods, as is the case for example in the 500 MW Telanok Scheme in the Cameron Highlands of Malaysia, variations in discharge of considerable magnitude can occur over short timescales, creating artificial freshets or floods downstream.

Links between river regime and geomorphological processes are complex. Sediment is often deposited in the relatively still waters of impoundments, often leading to considerable problems of loss of storage capacity. Allen (1973) presents a post-construction analysis of the Anchicaya Hydroelectric project in Columbia highlighting sedimentation problems, among many others. Reservoir siltation is particularly a problem in river basins with high rates of subaerial erosion and sediment transport, such as China (Yuqian and Qishun 1981). Khogali (1982) describes the problem of siltation in the Khashm el Girba Reservoir on the Atbara River in Sudan. This supplies water to an irrigation scheme on which Nubian evacuees from the Aswan Dam were resettled. River discharge is highly seasonal, and the reservoir is filled at peak flood season when the sediment load of incoming water is high. Silt has accumulated at 50-55 million cubic metres a year, and by 1977 the original storage capacity had dropped by 59 per cent. It is estimated that it will drop by a further 38 per cent by 1997, and that by 2025 there will be little water for irrigation. Irrigation efficiency and extent are already affected (Khogali 1982). Possible solutions include more efficient irrigation water use, and the construction of two upstream dams simply to act as silt traps. These might have additional power generation or irrigation functions, but obviously drastically alter the original economic justification of the project.

Sediment deposition within reservoirs can lead to clear water releases below the dam. Rasid (1979) reports sediment loads reduced by 91 per cent following dam construction on the South Saskatchewan River. The corollary of reservoir sedimentation is therefore very often erosion immediately downstream of the dam (see for example Park 1977, Petts 1977, 1980). However, further downstream such effects become more complex (Simons and Li 1982, Kellerhals 1982, Williams and Wolman 1984); thus Graf (1980) for example records stabilization of rapids downstream of the Flaming Gorge Dam on the Green River in Utah.

There are relatively few studies of downstream degradation following dam construction in the Third World, although it is clear that this lack of research does not reflect the importance of the problem. Olofin (1984) discusses downstream channel change in northern Nigeria, comparing channel form below the Tiga Dam on the Kano River (closed 1973) with that on the then uncontrolled Challawa River (a dam is currently under construction at the Challawa Gorge, West of Kano). The channel of the Kano River appears to have become stabilized as a result of control, changing from an ephemeral to a vegetated perennial channel, with resulting impacts (on balance favourable) on land use because of the possibility of year-round cultivation in and adjacent to the river bed.

Attwell (1970) and Guy (1981) describe river bank erosion on the Zambezi downstream of the Kariba Dam, in the Mana Pools Game Reserve. Many of the river banks in this area are of relatively erodible alluvial soils and unconsolidated sands, and a certain amount of erosion would be expected under 'normal' conditions. However, he argues that releases from Kariba have caused increased bank erosion, and estimates that 230 hectares of land have been lost to the river between 1965 and 1973, 26 hectares per annum. He blames unseasonal high flows, high flow levels, rapid fluctuations in flow levels, and the fact that water released by the dam is low in sediment for this increased erosion, and argues that even though erosion (particularly of the weaker bank substrates) certainly occurred before dam construction, this was slow enough to allow the establishment of *Acacia albida* woodland. He suggests that the dam has hastened normal river system dynamics.

The impacts of dams on river channels has been recognized by engineers for some time, and there are a number of studies of tropical dams attempting to predict the nature and severity of such effects, for example Kerssens and Zwaard (1980) on the Tana River in Kenya, SWECO (1978) on the Ruaha in Tanzania and Pickup (1980) on the Purari River in New Guinea. Pickup models the response of the Purari following construction of a hydroelectric dam at Wabo. The Purari River flows south from the central highlands of New Guinea. Its basin is mountainous (up to 4,000 m), and at the Wabo Dam site covers over 26,000 square kilometres. Rainfall is high, between 2,000 and 8,000 mm, and river discharge is large (average 2360 cubic metres per second) and highly variable (487-10,450 cubic metres per second instantaneous maximum and minimum discharges). The sediment load is vast, some 57 million tonnes a year (Pickup 1980). Downstream of the dam site, the river channel is between 200 and 400 metres wide, its bed material coarse gravel becoming steadily finer towards the sea.

Storage requirements on the Purari River are limited, so the purpose of the Wabo Dam is to create a guaranteed head for power generation, involving a reservoir of 290 square kilometres and 16.6 billion cubic metres of storage. Research involved the collection of data on river discharge and sediment yields, and the use of computer simulation of reservoir operation. The disturbances of the construction period (largely the addition of sediment to the river) were too complex to model, and the filling period was expected to be short (two to three months), and so was not modelled. Simulation of subsequent changes over the first two years following closure, however, suggested that there would be relatively rapid adjustment of the downstream channel following the closure of the dam, until the creation of a new equilibrium with relatively slow changes in sediment balance. Initially, the volume of sand-sized material reaching the delta would be reduced by 63 per cent, silts by 45 per cent and clays by 18 per cent. This would be likely to be accompanied by extensive bank and some bed erosion. In the longer term, little except clay particles would pass the dam in its first 200-300 years, and sand input to the delta would be reduced by 78 per cent, silt 53 per cent and clay 22 per cent. However, it was predicted that channel degradation might be limited by the armouring of the bed by coarse material, even in the sensitive stretch immediately below the dam.

In this it might be said that the engineers were fortunate, for such armouring by no means always occurs. Kabuzya *et al.* (1980) assessed the downstream impact of the proposed dam at Stiegler's Gorge in Tanzania, and concluded that degradation downstream of the dam would be very rapid because river bed sediments are fine and there are no rock beds or other constraints on degradation. In mitigation they suggest a series of barrages downstream, or stretches of bed artificially armoured with rocks. Neither are attractive prospects in economic terms.

The physical impacts of dam construction extend downstream into delta and estuary environments. The significance of such effects has been argued in a number of cases, notably that of the Nile Delta following the closure of the Aswan High Dam. The original dam was built in 1902 and raised in 1912 and 1933. Engineering studies began for the new dam in 1953-4, and it was completed in 1969 at a cost of $625 million (Farid 1975). Planned storage included silt trapping capacity sufficient for an estimated 500 years. Questioning of the wisdom of the new development at Aswan began early (see for example van der Schalie 1960). Against gains which included hydroelectric power, increased water availability (which allowed the expansion of the irrigated area in Egypt and extensive conversion to double cropping) and flood protection, it was argued that there were problems of reduced soil fertility in the Nile Valley because of the lack of sediment in floodwaters and consequent erosion of the delta, and reduced flows which led to saline penetration of coastal aquifers (Farid 1975, Biswas 1980, Shalash 1983). Talling (1980) argues that the data necessary to assess the changed seasonal pattern of discharge following the construction of the high dam have not been published. Nonetheless, various studies suggest that erosion in the Nile Delta caused ultimately by the restriction of sediment supply by the High Dam is a serious problem, particularly for fishing villages and highly productive coastal lagoon fisheries (Kassas 1973, Sharaf el Din 1977). Further links with marine fisheries in the eastern Mediterranean were also made by George (1973). The existence of erosion in the delta following the construction of the dam is confirmed by Murray *et al.* (1981). Bathymetric surveys offshore identified a new sand ridge system acting as a sink for eroded material.

Physical impacts downstream of dams are clearly both common and highly complex. Furthermore, they can extend for many hundreds of kilometres downstream, and well beyond the confines of the river channel. Barrow (1987), for example, discusses the possible impacts of the Tucurui Dam on the Tocantins River in Brazil on riparian cultivation on the *varzea* land because of reduced inputs of silt. Physical and hydrological impacts in their turn influence the nature and severity of change in riverine and floodplain ecology.

The ecological impacts of river control

The most direct ecological effect of dam construction is obviously the loss of upstream terrestrial environments. Singh *et al.* (1984) describe the significance of the loss of 530 hectares of species-rich rainforest in the Silent Valley Forest reserve on the Nilgiri Plateau, India, and the attendant disturbance of surviving forest during project construction. Beyond this crude and direct impact, however, there is a series of other more complex and sometimes subtle effects of dam construction on the ecosystems of the controlled river.

There is a considerable body of knowledge on the ecology of running waters (Hynes 1970, Whitton 1975, Lock and Williams 1981, Davies and Walker 1986). Transformation or modification of discharge patterns and stream environments have a range of significant effects on those ecosystems. The most obvious of these concern the ecological succession of the newly-created impoundment as the organisms of flowing water (lotic ecosystems) are replaced by those of still (lentic) ecosystems, and planktonic and littoral species arrive (Baxter 1977). The limnology of reservoirs is distinctive, and the evolution of tropical impoundments in particular has been of considerable interest to ecologists (Goldman 1976). In 1969, White argued that research effort had been on too small a scale and had started too late to provide 'either useful predictions or firm conclusions, about the course of biological events consequent upon impounding water in the Tropics' (White 1969, p. 37). To an extent this complaint is still valid, despite two decades of research, because of the complexity of succession in tropical ecosystems.

The development of the reservoir ecosystem depends partly on the chemistry, turbidity and temperature of inflowing waters, and partly on the nature of the substrate inundated. At Kainji on the Niger, 37,000 hectares of vegetation were cleared and burned. A diverse benthic fauna developed. The Volta Lake behind the Akosombo Dam, on the other hand, filled much more slowly (over a seven-year period), and there was no clearance of vegetation. As a result, there was initially considerable deoxygenation at depth as vegetation decayed. The lake became poor in nutrients with a low crop of phytoplankton (Baxter 1977).

In neither of these cases was there a severe infestation of algae or macrophytes of the kind that has occurred elsewhere, for example in Cabora Bassa or Kariba on the Zambezi (Balon and Coche 1974, Davies *et al.* 1972). Floating plants such as the Nile Cabbage (*Pistia stratiotes*) or the water hyacinth (*Eichornia crassipes*) can create problems for hydroelectric turbines, and increase evapotranspiration losses from the reservoir surface. On the other hand they can, to an extent, provide substrates where fish can find food sources. The establishment of reservoir fisheries in the Third World also depends on the way in which the reservoir ecosystem develops, and there are no simple rules. The development of the fish fauna in tropical reservoirs is complex (Lowe-McConnell 1985), and may involve an initial peak of population as nutrients from flooded areas feed into the ecosystem, followed by a slump to a lower level (Jackson 1966, Petr 1975). Calculations of potential catches from new reservoirs based on data from small (and often much shallower) natural lakes are rarely of any value. Furthermore, reservoir fish stocks will be at least in part pelagic, and may require very different fishing techniques to those used by previous fishing inhabitants of a river floodplain.

The impacts of dam construction on running water ecosystems are relatively well understood in temperate rivers, particularly where there are important sport fisheries. The impacts of dams on salmonids, for example, were studied in both Britain and North America in the 1960s and 1970s (Raymond 1968, Alabaster 1970, Davidson *et al.* 1973). The result of this work is a relatively detailed understanding of the migratory patterns and in-stream habitat requirements of salmonids, and the development of responses in dam design, dam operation and downstream river management to minimize adverse impacts of control on fish stocks (Hayes 1953, Swales and O'Hara 1980). Knowledge of tropical rivers is less complete (but see Payne 1986), although there is a considerable amount of data on freshwater fisheries, particularly in Africa (Hickling 1961, Lowe-McConnell 1975, 1985, Welcomme 1979), and the limnology of certain rivers such as the Nile (Rzoska 1976) and others in tropical Africa (Davies 1979).

Passage through a reservoir has a number of effects on water quality. Water released from low outlets in a dam may be deoxygenated, and sometimes rich in hydrogen sulphide, and it may also be cold. There may be influences on invertebrate drift, and river bed degradation downstream can lead to the loss of important in-stream spawning grounds. In tropical floodplain rivers, however, it is the impact of dams on natural flood regimes which is the most significant. Many fish exhibit fairly short 'lateral' migrations or longitudinal migrations of greater length in response to seasonal fluctuations in the river. As water spills out onto the floodplain it becomes enriched with organic matter from decaying vegetation, animal manure and other materials. This creates a flush of algal, bacterial and zooplanktonic growth, which forms a food source for invertebrates. Aquatic and emergent vegetation growth is also rapid and extensive. As a result there is abundant food for fish which follow the floodwaters out of the river channel, and in particular both food and shelter for young fish. Growth rates are very rapid in these flood conditions. As the flood subsides, fish move back to the river channel, and in many cases eventually to the small and deoxygenated pools of largely dry river beds. This season sees high mortality, both from fish stranded in evaporating floodplain pools and those taken by birds, predatory fish and human predators (Lowe-McConnell 1985).

Dam construction has significant implications for the fish population of tropical floodplain rivers. Jubb (1972) records the failure of *Hydrocynus vittatus* and other species to spawn in the Pongolo floodplain in South Africa following river control which reduced the annual flood to such an extent that floodplain pools remained isolated from the main channel. The Kainji Dam acts as a complete barrier to fish movement, and although catches at the foot of the dam are high, studies further downstream reported significant reductions in fish catches between 1967 and 1969 (Lelek and El-Zarka 1973, Adeniyi 1973, Lowe-McConnell 1985), and associated reductions in fishing activity. Similar reductions in fish catches and fishing effort below dams were recorded elsewhere in Nigeria, for example on the Sokoto (Adams 1985a). A more complex response was shown by the *Egeria* clam fishery of the lower Volta. Breeding in the clams is triggered by a rise in salinity following reduced river flow. During construction of the dam at Akosombo the critical salinity conditions moved 30-50 km inland, and once operating the constant flows pushed the fishery down to within 10 km of the river mouth (Lawson 1963, Hilton and Kuwo-Tsri 1970, Chisholm and Grove 1985). The further impacts of reduced flows in the Volta in the mid-1980s on the clam fishery are not known, nor is the importance (for better or worse) of the dams at Akosombo and Kpong.

Floodplain vegetation communities, both those immediately bounding river channels and those of larger and more extensive river-fed wetlands, are influenced by flooding patterns in much the same way as aquatic ecosystems. Among the most extensive studies of such impacts are those of the University of Zambia under the Kafue Basin Research Project, begun in 1967 (Williams and Howard 1976, Howard and Williams 1982, Handlos and Williams 1984). The development of a hydroelectric dam in the Kafue Gorge was mooted in the 1950s at the time of the Central African Federation, and would in many ways have been more logical than the Kariba project which was eventually adopted (Williams 1977). In 1967, eleven months after UDI, Zambia announced that development would go ahead, and work began on the Kafue Gorge Dam in 1967. It was completed in 1972, and the second phase of the project, further generation capacity in the Kafue Gorge Dam and a second dam at Itezhitezhi above the Kafue Flats was completed in 1982. There has been considerable concern about the environmental impacts of the developments at Kafue, particularly concerning the Kafue Lechwe (Rees 1978a, b), but although downstream areas of the flats have been flooded by the Kafue Gorge Dam, maximum river levels have not changed significantly. While concern continues (Sheppe 1985), it seems that environmental impacts have not been as serious as some had feared.

In semi-arid Africa, the river floodplains frequently support woodland vegetation in areas of dry savanna bush (Hughes 1987). These riparian or riverine woodlands are supported by high groundwater tables fed by river flows, and the outer edge of the forest is determined in part at least by the depth to water table. However, forest regeneration depends on periodic high flood flows. The maintenance of the forest therefore depends in a complex way on the annual and inter-annual pattern of river discharge. Alterations in that discharge pattern, for example from dam construction, can have significant impacts on the viability of the forest ecosystem (Hughes 1987).

The implications of such river control are understood for few rivers or wetland areas, but have been studied in detail on the Tana River in Kenya (Hughes 1984). The Tana River flows for about 650 km from the slopes of Mount Kenya to enter the Indian Ocean north of Malindi. A series of dams has been built in the headwaters, notably at Kindaruma in 1968, Kamburu in1975, Gitaru in 1978 and Masinga in 1982. Below these the river flows across a plain of low elevation with Sahelian bush savanna vegetation. Rainfall is low, rising from 300 mm to 1,000 mm at the coast. Groundwater recharge by the river, particularly at high flow stages, supports a narrow belt of forest 1-2 km wide. This now exists in a series of discontinuous blocks, partly because of clearance by Pokomo and Malekote people who live along the river. Some of these blocks support two endemic primates, the Tana River Red Colobus (*Colobus badius badius*) and the Tana Mangabey (*Cercocebus galeritus galeritus*), which have received protection in the Tana River Primate reserve.

The floodplain forest is diverse in structure and floristics, reflecting the complex geomorphology of the floodplain itself (Hughes 1984, 1987). Evergreen species occur on heavier soils, trees such as figs (*Ficus sycomorus*) are important on sandy levees, there is an endemic poplar (*Populus ilicifolia*) which forms a succession on pointbars, and oxbows exhibit a complex succession of low thorny scrub vegetation.

Drier parts of the forest are dominated by species of Acacia. The dynamics of succession in the forest are complex, but the importance of river flooding patterns is clear. The Tana floods twice a year, with low flow periods between February and March and between September and October. There are periodic high flows, particularly in May, which inundate extensive areas of the floodplain. There is considerable inter-annual variation in flows, and the record shows high flood years every few decades. Thus high flows were recorded in 1961 which were three times the average annual maximum flow (2,418 cubic metres per second). Although this event seems to have had no effect on forest regeneration, it seems likely from studies of tree girths and growth rates that past regeneration has been associated with extreme flows of this kind. Further work on the impacts of the dam at Masinga suggests that it will reduce both the height and frequency of high flows in the lower Tana. It is possible that this will bring forest regeneration to an end (Hughes 1984). Other pressures of development local to the forests, notably the cutting of construction timber and fuelwood for the irrigation scheme at Bura, also appear to be having serious impacts on the forests of the lower Tana (Hughes 1984, 1987).

Tropical floodplain ecosystems are closely adapted to existing patterns of discharge, and are hence vulnerable to environmental change caused by dam construction. However they are also both ecologically productive and economically important environments. Impacts on natural processes, particularly the ecology of fish production, can therefore have significant socio-economic implications. The intensity of human use, and hence the potential severity of socio-economic impact are particularly great in extensive floodplain wetlands, for example those of arid or semi-arid Africa.

Ecology and development of tropical wetlands

In ecological terms, the environmental effects of development projects are broadly similar in the First and Third Worlds. A dam transforms the ecology of a river by changing patterns of downstream discharge in more or less the same way in Teesdale in the North Pennines of Britain as the Niger Valley. What differs is the significance of the resulting effects. In Teesdale the Cow Green Reservoir, built in the 1960s, had a relatively minor (and not entirely injurious) impact on populations of the brown trout (Crisp *et al.* 1983). The Kainji Dam on the other hand had a major effect on artisanal fishery of the River Niger, with serious secondary socio-economic impacts (Adeniyi 1973). Environmental concern in the industrialized world has focused on environmental quality, or amenity. Indeed environmentalism in Europe particularly has been criticized precisely because of the inherent elitism of this focus (see Chapter 2). Third World environmentalism, although embracing preservation and wildlife conservation (e.g. Singh *et al.* 1984), perforce adopts a distinct focus, on life and livelihood (Adams 1982).

The environmental impacts of development in the Tropics have a particularly immediate and important influence on human welfare because of the extent to which subsistence depends on natural or semi-natural ecosystems. This dependence of subsistence on ecology is well-demonstrated by the example of tropical wetlands, particularly those which exist in semi-arid regions such as sub-Saharan Africa. Tropical wetlands are among the most ecologically productive of global ecosystems and in the drier Tropics in particular they can be of enormous socio-economic importance. They are also important in other ways. The wetlands of the Sahel, for example, such as the Inland Delta of the River Niger in Mali and Lake Chad are important staging-posts for Palaearctic birds on migration to and from wintering grounds in tropical Africa. These 'functions' of wetlands, as the IUCN Wetlands Conservation Strategy terms them, embracing flood control, food production and wildlife conservation, are sustained by hydrological and ecological processes. They are therefore threatened by developments which alter hydrological patterns and transform wetland ecosystems.

Many such wetlands have become the subject of interest from development agencies. Many others have become inadvertantly influenced by development projects elsewhere, particularly in upstream environments. Thus in the Okavango Delta in Botswana (Botswana Society 1976) there are now investigations of the feasibility of harnessing water initially for water supply to diamond mines further south by reversing the flow in one of the rivers draining the wetlands, keeping Lake Ngami dry and forming a reservoir around the town of Maun. Subsequent possibilities include the diversion of water from the delta itself for use in irrigation development.

Developments affecting the swamps of the Sudd in the Sudan have reached a more advanced stage (Tahir 1980, Howell 1983, Howell *et al.* 1988). The perceived problem here is the volume of water lost from the White Nile in the swamps of the Sudd, and the context is the continuously rising demands for water by Egypt. Division of the Nile's flow between Egypt and the Sudan was determined by the Nile Waters Agreement of 1929, which was revised in 1959 (Godana 1985). This allocates Egypt 55.5 million cubic metres against Sudan's 18.5 million cubic metres. The possibility of bypassing the Sudd was investigated exhaustively in the 1940s and 1950s by the Jonglei Investigation Team, who reported in 1954. The Equatorial Nile Project as then conceived had significant adverse environmental impacts (Howell 1953), and with the building of the High Dam at Aswan was not pursued.

However, the canal was proposed again in 1974 in a revised form. Construction began, but was halted by the Sudanese civil war in the early 1980s. The canal, being constructed from south to north, is 54 m wide and will be 360 km long. It will carry 20 million cubic metres of water per day. This is smaller than the original scheme. Extensive adverse environmental impacts were feared (el Moghraby 1982), but surveys revealed that owing to a fortuitous and unexplained increase in the discharge of the White Nile in the 1960s, impacts on the Sudd will be less than was at one time feared. If high flows in the White Nile persist (and there is some evidence in the late 1980s that they will not), it is expected that the area of seasonally flooded *toich* grassland will be reduced by 25 per cent, and there will be impacts on fish populations. There will also be serious problems of access across the canal for Dinka pastoralists (Howell 1983, Howell *et al.* 1988, Moghraby and el Sammani 1985). In a sense, however, the direct environmental impacts of the Jonglei project, even if limited, are but one part of the wider intervention by the Sudanese state in the area. The introduction of irrigation and ranching, and the improvement of road communications with cities in the north of Sudan will also cause a significant, and not universally positive, impact on the economic and social bases of Dinka society (Lako 1985).

Where floodplain ecosystems are intensively used by people, impacts of development can be more direct and immediate. In the Sokoto Valley of northern Nigeria, as in the floodplains of many West African rivers, agriculture has long been practised on seasonally flooded land by the Hausa. The Sokoto is, like the Inland Delta of the Niger, one of the centres of domestication of the endemic West African rice, *Oryza glaberrima*, and flood cultivation is centred on both this and Asian rice (*Oryza sativa*) which was introduced probably 300-400 years ago. Floodplain cultivation is integrated with upland rainfed cultivation. There is a single rainy season, from about the end of May to the end of September, and the River Sokoto has a strongly seasonal regime, with high peak flows in July, August and September. Rice and flood-resistant sorghum is planted in the floodplain in the rains, with rainfed millet and sorghum followed by relay crops of cotton, cowpeas or groundnuts on upland areas. There are a great many local varieties of rice and sorghum in the floodplain, adapted to particular conditions of flooding, soil waterlogging and desiccation. About 90 per cent of the floodplain is cultivated. Both rainfed and floodplain crops are harvested from October onwards, and the floodplain then comes into its own because a second crop can be grown using residual soil moisture and sometimes shallow groundwater irrigation. This cultivation is highly variable, but often includes the cultivation of peppers of various kinds, onions and sweet potatoes.

In 1978 the Bakolori Dam was completed on the Sokoto River, and in subsequent years it brought about a significant reduction in peak flows in the Sokoto (Adams 1985a). This in turn reduced the depth, duration and extent of inundation in the floodplain for 120 km downstream before the next major confluence with the River Rima. In three survey villages, the average reduction was about 50 per cent in the extent of flooding.There was as a result a reduction in the area cropped (from 82 per cent to 53 per cent of plots in one village), and a particularly marked fall in the area under rice (from about 60 per cent of fields to 14 per cent in one survey village) and in the amount of dry-season farming. The proportion of households undertaking dry season cultivation fell from 100 per cent to 27 per cent in one village (Adams 1985a). It was estimated that of a total of 19,000 hectares of floodplain land, the dam caused the loss of 7,000 hectares of rice and 5,000 hectares of dry season crops.

To an extent these losses were compensated for by increases in the area under millet and sorghum, but the new uncertainty about flooding patterns and the intolerance of millet to waterlogging following rain on heavy soils meant that farmers were not able to adapt wholly to the new conditions. Furthermore, although irrigation was a known and tested technology in the Sokoto Valley, the costs of well digging and the labour demands of water lifting were both increased because reduced floods were accompanied by increased depth to water table. As a result irrigation became a more specialized technique, only accessible to larger producers (Adams 1988b). More recently, the Sokoto valley has shared with other parts of northern Nigeria in a boom in small petrol pumps for irrigation, and it is possible that the economies of production have changed. Nonetheless, the initial impacts of the dam itself remain clear.

The magnitude of lost production in the Sokoto floodplain needs to be seen against predictions that the flood-control effects of the dam would allow increased production of rice from downstream areas. The value of lost downstream production can be estimated, and shown to have a significant effect on the benefit/cost ratio of the Bakolori Project as a whole (Adams 1985a). Clearly environmental impacts not only impinge on the livelihoods of the poor, they can be highly relevant to the cost calculations of developers as well. For this reason, environmental impacts ought to be a major concern of project planners. The nature and success of attempts to incorporate the environment into project appraisal is discussed in the next chapter.

7

Environment and Development Planning

Reforming development planning: technocratic responses

Since the 1960s there has been a growing understanding of the adverse ecological impacts of development. One result of this has been the attempt to identify specific formulae for avoiding or minimizing these impacts. Such an approach is essentially one of technocentrist environmentalism, looking to reform the development process to optimize economic returns and environmental impacts. This search for 'environmentally benign' development has been a central feature of sustainable development thinking. It became explicit in the work of the International Union for Conservation of Nature (IUCN) and the International Biological Programme (IBP) and the Man and the Biosphere Programme (MAB) in the 1960s and early 1970s (See Chapter 2).

A conference on 'the ecological aspects of international development' was held late in 1968 at Airlie House in Virginia, organized by the Conservation Foundation and Washington University of St Louis. Its proceedings were published as *The Careless Technology* (Farvar and Milton 1973). A series of meetings followed this between conservation organizations (IUCN and the Conservation Foundation), development agencies (notably UNDP, UNESCO and FAO) and the International Biological Programme (IBP). In 1970 there was a meeting at the FAO headquarters in Rome of interested parties, including now USAID and the Canadian aid agency CIDA, to consider further the ecological impacts of development. As a result of this it was decided that IUCN and the Conservation Foundation should publish guidelines for development planners. These appeared in 1975 as *Ecological Principles for Economic Development*. Initially the book was to include discussion of 'interrelationships between economic development, conservation and ecology', but in the event it was restricted to an exploration of ecological concepts 'useful in the context of development activities'. Particular emphasis was placed on tropical rainforests and semi-arid grazing lands, the effects of tourism (particularly in fragile environments such as high mountains and coasts) and the development of agriculture and river basins (Dasmann *et al.*. 1973, p. vi).

The message of the book is, in retrospect, very familiar. Many of the concerns rehearsed in the previous two chapters of this book are found there, for example ideas of carrying capacity, overgrazing, soil erosion, desertification and pesticides. This familiarity is hardly surprising in that it encapsulated post-war thinking by conservationists and ecologists on development, and it has formed the basis of the subsequent decade and a half of thinking and writing about sustainable development. Its basis is the application of concepts from the science of ecology to development activities. The central argument is that ecology has a direct bearing on what can, or should, be done in development. If the 'lessons' of ecology are ignored, 'entirely unexpected consequences can often result from what are intended to be straightforwardly beneficial activities'. Thus for example, the replacement of tropical rainforest with a palm oil plantation 'sets in motion complex ecological forces', which may involve the loss of equilibrium and pest outbreaks (Dasmann *et al.* 1973, p. 44).

Dasmann *et al.* picture ecology as an integrative science, an interdisciplinary way of thinking which could be instilled in the minds of 'forester, agricultural specialist, range manager ... development economist or engineer'. The use of ecology in development planning has the aim of both 'enhancing the goals of development' and 'anticipating the effects of development activities on the natural resources and processes of the larger environment'. Ecology had in the past been confined to assessing the potential productivity of a resource; now the adverse impacts of 'certain kinds of development and management technology' on local and global environments demanded that these impacts be 'anticipated in the planning process, and to the extent possible, evaluated', so that decisions about developments are made 'in full knowledge of possible consequences'. A decision to develop despite environmental impacts might be justified by counterbalancing benefits, but 'it should never be taken blindly' (Dasmann *et al.* 1973, pp. 21-2).

Ecological Principles for Economic Development was a pragmatic attempt to express the views of environmentalists in a way that those responsible for decisions in development would understand and take account of. Adverse environmental impacts of development needed to be understood and dealt with during development planning if major problems were to be avoided. Ecology was therefore extremely useful.

Its application would help planners 'to make sure of success' (Dasmann *et al.* 1973, p. 21).

This pragmatic approach to development on the part of environmental organizations like the Conservation Foundation and IUCN was matched by interest from the technical disciplines they hoped to influence. This is shown most clearly in the response to the question of the environmental impacts of dams. The nature of these impacts, and rise of environmental concern in the 1960s and 1970s, are reviewed in Chapter 6. The apparent 'conversion' of engineers to wider environmental concerns needs to be seen in the context of continuing suspicion, hostility and pressure from environmentalists (e.g. Morgan 1971, Drew 1976).

By the 1960s large dams were sufficiently common for it to be worth the construction industry compiling a global register (Brown *et al.* 1964), and by the 1970s engineers were starting to acknowledge and review environmental effects (Turner 1971). By the second half of the 1970s, such reviews were commonplace in journals and magazines read by engineers and hydrologists (e.g. Biswas and Biswas 1976, writing in *Water Power and Dam Construction*), and it was possible to synthesize more than a decade of ecological research (Baxter 1977). The Man and Biosphere Programme designated Project 10 on the effects on man and his environment of major engineering works (UNESCO 1976). In 1977 the UN Water Conference received a report on *Large Dams and the Environment: Recommendations for Development Planning* (Freeman 1977). The International Commission on Large Dams (ICOLD) appointed a committee on damming and the environment in 1972, that in 1981 published *Dam Projects and Environmental Success*, a paper 'intended to illustrate the concern and knowledge of dam engineers related to environmental matters' (ICOLD 1981, p. 8). This paper, with its review of impacts, its stress on the fact that environmental effects could be beneficial as well as adverse, and that 'dam engineers are very aware of the importance of environmental requirements and give full and early consideration to possible difficulties of this type' (p. 7) neatly combined professional training and public image creation. It also demonstrates the success, at least on a rhetorical level, of attempts to imbue development with ecological awareness.

Environmental Impact Assessment

Such attempts were further developed through the formulation of principles of Environmental Impact Assessment (EIA) in the late 1960s. EIA became an important part of public policy in the industrialized world with the US National Environmental Policy Act

(NEPA) of 1969. A number of industrialized countries have followed to a greater or lesser degree down the path of institutionalizing the assessment of environmental impacts (Park 1986). One of the international applications of EIA procedures was again by ICOLD, who designed and tested a version of the Leopold Matrix (Leopold *et al.* 1971) to take account of the impacts of large dam construction (ICOLD 1980).

The procedure of EIA was taken up in the 1970s by the Scientific Committee for Problems of the Environment (SCOPE). SCOPE was set up in 1969 under the International Council of Scientific Unions following General Assembly of the International Biological Programme (IBP) in Bulgaria. It is one of eleven scientific committees of ICSU, and has representatives of thirty-four member countries and fifteen unions and scientific committees. Its mandate was

- to assemble, review and assess the information available on man-made environmental changes and the effects of these changes on man;
- to assess and evaluate the methodologies of measurement of environmental parameters;
- to provide an intelligence service on current research;
- by the recruitment of the best available scientific information and constructive thinking to establish itself as a corpus of informed advice for the benefit of centres of fundamental research and of organizations and agencies operationally engaged in studies of the environment.

(Munn 1979)

SCOPE published a series of monographs in furtherance of these aims, a number of which are relevant to the environmental impacts in the Third World. A number of early reports were global in range, addressing, for example, global environmental monitoring (1971) and the proposed Global Environmental Monitoring System (GEMS) (1973). In 1974 they published the proceedings of the SCOPE/UNEP symposium on 'environmental sciences in developing countries' in Nairobi (SCOPE 4). The second SCOPE report in 1972 concerned man-made lakes as modified ecosystems, and in 1975 they published a volume reviewing techniques in environmental impact assessment, *Environmental Impact Assessment: Principles and Procedures*. This was subsequently revised at a meeting in 1977 and republished (Munn 1979).

EIA procedures are diverse, and have been developed in widely different policy contexts (O'Riordan and Hey 1976, Clark 1976, Munn 1979, Chapman 1981, Park 1983, Lee 1983). Munn (1979) suggests that an EIA should:

- describe the proposed action, and alternatives;
- describe the nature and magnitudes of likely environmental changes;
- identify the relevant human concerns;
- define criteria to be used in measuring the significance of environmental changes, including relative weighting given to different changes;
- estimate the significance of predicted environmental changes;
- recommend acceptance or rejection of the project; any necessary remedial action or alternatives;
- recommend inspection procedures to be followed after the action is completed.

These aims are close to those specified by the US legislation. NEPA stipulated that all major federal initiatives should be accompanied by an Environmental Impact Statement (EIS), which gave account of the environmental impacts of the action, considering both long and short-term impacts, issues of resource commitment implied by the development, and alternatives. These generalized aims were expanded by a series of guidelines produced by the Council for Environmental Quality and other organizations in the USA (O'Riordan 1981, Chapman 1981).

The methods of EIA are manifold. Most demand procedures to tackle a series of problems, from the identification of the nature of impacts, the prediction of their severity, the interpretation of impacts (for example the comparison of different impacts), and the communication of the results (Munn 1979). The starting point is often the establishment of checklists of potential impacts, but these are often unable to take account of secondary and tertiary impacts. One response to this is the use of a matrix approach such as that developed by the US Geological Survey (Leopold *et al.* 1971). Project actions and possible environmental impacts are placed on one axis of a matrix and conditions and characteristics of the environment on the other. The boxes of the matrix can then be filled, either using simple binary codes or evaluations of the magnitude (scale or extensiveness) and importance of the impact.

There have been many analyses and critiques of EIA procedures, and suggestions for reform (O'Riordan 1976a, Munn 1979, Lee 1982, Wathern 1988). One obvious problem concerns quantification. Even at the stage of assessing the severity of environmental impacts, the science of ecological survey starts to give way to informed value judgement of the analyst. These problems are compounded at later stages of the assessment in the interpretation of impacts, although there are a series of methodologies for determining weights or scores in a quasi-independent manner, for example that of the Batelle Columbus Laboratories (Dee *et al.* 1972). Quantified evaluations, like those of cost/benefit analysis, may be criticized because they make evaluation the preserve of the 'expert', remote from public comprehension and accountability. Whether cloaked in quantification or not, such procedures are essentially qualitative, and therefore highly dependent on the skills, prejudices and perceptions of the analyst. Quantified methods may appear to confer enhanced legitimacy, particularly among a scientific audience, and may also make the task of making decisions on the basis of the EIA easier by effectively pre-judging the issue. They are therefore attractive both to scientists and consultants wanting to present a clear case, and politicians faced with difficult decisions. At the same time, however, they place excessive reliance on the skills of 'experts' often far removed from appointed arenas of decision-making. Quantified methods can also make the communication of results harder, particularly to affected groups (Lee 1983).

Another important criticism is the apparent lack of integration of EIAs into project planning procedures. O'Riordan (1976) points out the confused and ineffectual way in which EIAs are used as devices to tack environmental issues onto existing cost/benefit analyses. It is clear to most observers, albeit not always to economists, that not everything about a development project can be reduced to quantifiable terms. There are many intangible aspects of the impacts of project development, often including the environment (if only because of ignorance about impacts), and society and culture (Davos 1977). O'Riordan argues that there is a tendency to use EIA procedures to collate these unquantifiable factors together in a systematic way, but without integrating them into the cost/benefit calculation. In as much as EIA procedures satisfy environmental concern without allowing that concern to be reflected in adapted project design, they are counter-productive: a waste of resources and a negation of their integrative and potentially transforming role.

Clearly EIAs are only as good as the policy frameworks within which they are carried out. The SCOPE handbook lays particular stress on the development of administrative procedures for EIA, and the sequencing of tasks and the nature of the players among whom the EIA will be created, assessed and acted on (Munn 1979). Procedures work through the definition of national development goals and the establishment of policy and programme activities before particular projects are considered. The next step is to determine whether particular developments will in fact have a significant impact, before formal EIA procedures begin. The tasks do not end until the EIA has been reviewed by the competent body, the project implemented and a post-project audit carried out (Munn 1979).

It need hardly be said that many of these steps are sensitive to critical decisions and pressures within and on bureaucratic systems, and also that in many cases some or all of these procedures are skipped. It is, for example, one thing to have an EIA commissioned and carried out, but quite another to integrate it into decision-making. The size of EIA documents can be out of all proportion to the capacity of agencies to digest data and reach informed conclusions. Indeed, the pursuit of gigantism in an EIA is one strategy open to the promoter of a major project who is hopeful of drowning objectors in a flood of indigestible data. In the USA, for example, EIA documents were too long, their language was too technical and complex, and they failed to summarize issues vital to decision-making (Lee 1983).

Even in the USA, EIA has remained a procedure fraught with policy and political pitfalls (Meyers 1976). While there has been pressure for the adoption of legislation analogous to that in the NEPA elsewhere in the industrialized world, notably in the EEC (Hammer 1976, Wood 1982, Briggs 1986), few even among environmentalists believe that this model is ideal. Precisely because of the nature of the policy and political environment in which it exists, it is clear that the extent to which EIA can transform the nature of development to minimize adverse environmental impacts is severely limited. Given this experience in the industrial heartland, there are obviously question marks over the usefulness of EIA procedures in the context of development planning in the Third World.

Environmental Impact Assessment in the Third World

The importance of carrying out assessments of the environmental impacts of projects in the Third World is clear. Lohani and Thanh outline the importance of the assessment of both long-term and short-term impacts of rural development in Southeast Asia, using simple checklist and more complex matrix approaches, to avoid 'irreparable' environmental damage (Lohani and Thanh 1980, p. 215). Pescod (1978) stresses the need to quantify damages and benefits of development so that a social choice can be made on the merits of projects. The EIA methodologies open to investigators are broadly similar to those in developed countries (Ahmed and Sammy 1985). USAID projects in the Third World should (in theory) be subject to the same EIA procedures as Federal actions in the domestic arena (Horberry 1988). However, the constraints on the effectiveness of EIA procedures in developed countries are greatly heightened in the Third World. A series of factors make Third World EIA a great deal more problematic.

One problem affecting Third World EIA relates generally to the nature of environmental impacts themselves. While it is possible to review the impacts of development projects, as was done in Chapter 6, such a synthesis is only possible in the form of generalization based on relatively few case studies. In practice it is often far from easy to identify the nature of environmental impacts at a particular point in space and time, and this is particularly true of the Third World (Pescod 1978). Significant impacts are often remote geographically from a project, for example the impacts of logging or dam construction on sedimentation or river degradation downstream. They may therefore be beyond the boundary of the specific development project, and therefore be unperceived by project developers. Such problems are compounded if, as in the case of downstream floodplain wetland environments, the place where the impacts occur is physically remote from centres of planning and decision making, difficult of access and of marginal importance politically. In the case of the Bakolori Project in Nigeria (Chapter 6), consideration of impacts in the wider river basin effectively stopped at the design stage, and all attention was focused on a relatively small project area which comprised only the dam and irrigation scheme (Adams 1985a, b). The parallel experience in the Gongola Basin saw project development preceding the basin-wide planning which could integrate development needs, potentials and impacts in the basin (Adams 1985b).

Environmental impacts can also be delayed, occurring some time after project development, and can be secondary or tertiary effects, for example ecological change following hydrological and geomorphological change. Indeed, relatively few geomorphological or ecological processes have instant responses, and most relate to each other in dynamic and complex ways. The possible impacts of the Aswan Dam on the loss of the fertility of land downstream and on delta erosion are obviously lagged in this way. Environmental impacts can also be increased by other developments or natural changes, which have synergistic effects on ecosystems. The most obvious example here is the impact of drought on the discharge of controlled rivers in the Sahel.

In the Hadejia river basin of northeast Nigeria, for example, there have been significant problems of reduced downstream flows to large floodplain wetlands and irrigation schemes (Stock 1978, Adams and Hollis 1989). While local people, and decision-makers in the downstream State of Borno, blame reduced flooding on the construction of dams upstream, it is in fact extremely difficult to disentangle the effects of low rainfall and of river control on the discharge patterns. There is no doubt that there is a serious problem, but it is less clear exactly what the impacts of the dams have been. The problem is confounded in this case by a series of relative unknowns. These include the complex nature of channel change in the wetlands, which has undoubtedly been important in denying water to certain areas and may itself have been affected by upstream damming, and the response of the recharge of the aquifer beneath the river to dam- and drought-reduced flows in the river (Adams and Hollis 1989).

Clearly the nature and extent of environmental impacts can present considerable difficulties to the investigator. These are exacerbated by the enormous data-hunger of EIA procedures. EIA requires a great deal of environmental, and particularly ecological, data. Assessment of impacts demands large data sets and long time-series, and such resources are unusual. The problem is simply another version of that which faces most forms of planning in the Third World, the task of 'planning without facts' (Stolper 1966). The problem is not simply the lack of data (although this is real enough), but a deeper ignorance. In development economics 'all too often it is quite unclear precisely what question should be asked; sometimes the question is asked wrongly, and it is by no means certain that answers always exist' (Stolper 1966, p. 9).

There are of course various technical tricks to extend short data sets, for example the work of hydrologists who correlate a short run of river discharge figures with a longer run of rainfall data and generate a synthesized data set offering a long time series. Such methods have been devised for temperate environments, and assumptions about the variability and lack of secular change in conditions valid there may be misleading and even dangerous in the Tropics. For example, the enormous spatial and temporal variability of rainfall in areas like the Sahel within and between years is now becoming clear (Trilsbach and Hulme 1984, Hulme 1987). There are severe risks in extrapolating from short time-series to derive longer-term averages. The variability of tropical climates in the current century is also well-established, and again there are serious difficulties with the use of short data sets, particularly those relative to periods such as the 1960s (wet in Africa) or 1970s (dry) in the design of projects or in attempts to predict their effects (Adams and Grove 1987).

The lack of environmental data in the Third World is matched by a lack of ecologists and other scientists. EIA also demands a high degree of knowledge of the dynamics of ecosystems, and an ability to be able to make realistic predictions about their response to stresses of various kinds. It therefore requires a great deal of ecological knowledge, and experienced ecologists. These are in short supply in most places, but the dearth is particularly acute in the Third World. There are few textbooks on tropical ecology directed at Third World audiences, although there are some creditable exceptions such as Deshmukh (1986), and in most Third World countries there are few ecologists or other environmental scientists with extensive knowledge and expertise. Such people exist, and their numbers are increasing, but in practice EIAs do suffer from the lack of indigenous skills.

The established response to the lack of indigenous expertise is to bring in outside 'experts' in the form of consultants. The importance of such outside influences on planning and development is considerable (Stone 1980, Chambers 1983) There may be many reasons for calling outside experts (Stutley 1980): the department in the host country may be empire-building, the civil servant may want a 'fall-guy', more time on his farm or access to foreign travel; the experts may be the only way to obtain resources in the form of aid, coming as part of a tied aid package. Whatever their justification, foreign consultants and planners are ubiquitous in the Third World. Information and the power to act flow outwards and downwards from the rich, centralized modern core of the world's developed countries to the periphery of the Third World (Chambers 1980), and the employment of expatriate consultants and technical companies is a significant feature of this flow. The problem is particularly acute in the case of what Chambers (1980, 1983) calls 'rural development tourism', 'the brief rural visit by an outside professional who sets out from an urban centre' (Chambers 1980, p. 98), but the difficulty goes deeper than that.

Foreign experts may command a legitimacy denied prophets in their own country derived from their presentation of technology and experience, but this expertise may be less valuable than it at first appears. Thus Winid (1981) argues that one of the reasons why developments in the Awash Valley of Ethiopia have not been more successful was the lack of knowledge on the part of the French consultants of the political, ecological, social, cultural, scientific and technical aspects of the region. This in turn was related to the fact that those in the field only had contact with the local elite, lacked supervision and failed to transfer technology. There are therefore serious pitfalls in a planning system built on the backs of foreign experts. Foreign companies offering to carry out EIAs without suitable in-house ecologists may well fail to produce an integrated understanding of environmental impacts. Even if temperate zone ecologists are brought in as consultants they may yet prove a poor (and costly) substitute for local expertise, perhaps glibly repeating dominant preconceptions and myths about tropical ecosystems (for example about overgrazing or desertification) which have misdirected development so often in the past.

The lack of ecological expertise matters more because of the disciplinary bias inherent in the planning process (Chambers 1983). Project appraisal is dominated by the 'hard' technical disciplines such as engineering, hydrology and agronomy, and by the most technocratic of the social sciences, economics. Most project feasibility studies are undertaken by companies dominated by these disciplines, or by consortia typically dominated by engineering companies. These are not, however, the disciplines relevant to EIA. The 'soft' disciplines like ecology, geography, anthropology or sociology central to EIA (and vital to social cost-benefit analysis) tend to be marginal to the planning process. Increasingly they have a place in project planning, but it is a small one. Sociology and ecology are typically slotted in with perhaps a one or two man-month input on a project where the total planning input is several man-years. In the FAO's appraisal of the soil and water resources of the Sokoto River basin in Nigeria, the socio-economic input amounted to only 10 per cent of the total man-months. That allowance certainly does not reflect the lack of important social and economic aspects of the development projects proposed. This is amply demonstrated by the fact that it was primarily social and economic problems which dogged the Bakolori Scheme which was eventually built as a result of the study (Wallace 1980, Andrae and Beckman 1986, Adams 1987, 1988a). EIAs based on whistle-stop tours by visiting experts are unlikely to be of great value.

Ecologists and other experts from 'soft' disciplines are rarely employed on the full time payroll of engineering companies. They are rarely, if ever, project managers. Furthermore, those who do lead appraisal teams usually lack the knowledge and experience necessary to know how to feed environmental data effectively into the planning process. It is clear that environmentally sound development demands that environmental factors be taken into account early in planning. EIA is only useful if it influences the nature, speed and extent of development. This demands multidisciplinary teams; project appraisal is usually done by teams rich in a very few technical skills, and they give a low priority to environmental expertise. Furthermore, such teams are rarely if ever led by those from 'soft' disciplines. Those who do lead them rarely even have the knowledge and experience to know what those disciplines require and can offer. They do not appreciate how EIAs need to be integrated into the planning process. However, the marginal position of environmental disciplines also reflects the institutional bias of that planning process. Just as in developed countries, EIA is too often seen as a way of paying lip service to environmental problems without solving them (O'Riordan 1976b). Inevitably, Third World EIA is something grafted in various often loose ways onto social cost/benefit analysis (Abel and Stocking 1981). Cost/benefit analysis (CBA) is the established procedure by which technical decisions about project development can be made. The techniques involved are widely known and reviewed (Sugden and Williams 1978, Clark *et al.* 1981, King 1967). CBA is systematic, based on principles that are widely comprehended (peoples' choices as revealed by surveys and markets) and it produces a neat result which can feed directly into the planning process. Debate about the shortcomings of CBA is considerable, focusing particularly on the problem of determining social values, and the impact of the the technical complexity of CBA on public participation (e.g. Stewart 1975, Garcia-Lamor 1984, Derman and Whiteford 1985). Nonetheless, Abelson argues that there is no viable alternative method available for the appraisal of environmental aspects of projects, claiming that it 'meets more of the requirements of the evaluation process and more nearly satisfies the other criteria than does any other evaluation method' (Abelson 1979, p. 57).

EIA is often seen as a discrete and often superfluous element within the project planning process. As a result it tends to be stuck on to the outside of an otherwise integrated planing process. Often this marginal position is revealed physically by the existence of a separate environmental report which is not related to the rest of the project documents. A bias of this kind arose in the FAO's appraisal of the land and water resources of the Sokoto river basin in northern Nigeria in 1969 (Adams 1987). That study was stimulated by fears of soil erosion and land degradation in the basin, yet the volume on land use and agronomy was vague, lacked conclusions, and was not referred to elsewhere in the seven-volume report. The appraisal team was dominated by engineers and hydrologists and soil scientists, and its studies were focused almost exclusively on the selection of dam sites and areas for irrigation (Adams 1987). There were no studies of soil erosion across the basin, and no recommendations of policy towards rainfed agriculture or pastoralism from which it has been assumed that erosion came. No ecologist was included in the team.

Environmental studies of development projects, like investigations of social impacts, are often begun too late. Hamnet (1970) laments the way in which as a project sociologist he could not initiate ideas, but simply react to problems created by technical proposals. Jaundiced practitioners sometimes see sociology as the task of selling unattractive projects to unwilling participants, and in the same way an EIA can seem to become part of the process of justifying a project rather than a genuine means of appraising it. EIAs carried out at a late stage can have little significant impact on project planning. This seems to have been the case for example with the rather magnificent appraisal of the environmental impacts of the Jonglei Scheme in the Sudan. This thorough appraisal (Howell 1983) was begun in 1979, several years after work on the canal began. Fortuitously, it predicted relatively minor adverse impacts of bypassing the Sudd, because of increased White Nile flows since the 1960s, but even if the predicted effects on Sudd ecosystems and people had been quite disastrous, it seems unlikely that substantial redesign of the project would have been possible, let alone its abandonment.

The marginal impact of EIA procedures on the nature of development also reflects the way in which EIA procedures raise the costs of project appraisal, and the relatively large amount of time required to carry them out successfully. The cost of EIA is considerable wherever it is done. The voluminous documentation required by US legislation in California for example absorbs roughly 1 per cent of project costs, and this is further increased if there are delays over decisions (Sandbach 1980). Data collection needs in the Third World are certainly no less than those in industrialized countries. Furthermore, like ecological data, neither policies nor decisions are necessarily easy to obtain when needed. EIA can therefore become a significant element in the cost of project appraisal. Such costs are sunk before it is known whether the project is technically, environmentally and economically viable, and will not be recouped if the project does not go ahead. Appraisal investigations, whether they involve EIA or not, demand work in a number of technical disciplines, and thus frequently demand studies from overseas consultants. Such consultancy services are costly, and often include a large foreign exchange component. Funds for this have to be raised from foreign loans, even if they come in the form of development aid, and the cost of debt servicing creates a strong incentive to minimize the costs of such appraisals. The EIA component, especially if costly, tends to be seen as secondary to technical and economic issues of project viability and can therefore be dispensed with on the grounds of cost.

Although there is no simple solution to the lack of adequate baseline data with which environmental impacts can be assessed, data collection as part of an EIA procedure can fill gaps in knowledge. However, such data collection takes time to complete. The completion of an EIA can therefore hold back the project appraisal process, and delay the moment when economic benefits can be realized. Very often tight project development schedules prevent the adoption of effective EIA procedures. EIA therefore not only raises the cost of sunk investment in project appraisal, but postpones the commencement of streams of possible benefits. There are obviously strong economic and financial incentives to truncate or speed up EIA procedures, or even omit them altogether.

Policy, politics and project planning

Shortcomings exist with EIA procedures in most contexts. In the Third World in particular, they are greatly exacerbated by the complex political *mileu* within which project appraisal takes place. Political pressures on national politicians combine with the sometimes counter-intuitive functioning of bureaucracies and the commercial interests of companies to create an arena of decision-making where the technical merits of a project are just one factor among many. In theory development projects undergo a strict and technically sophisticated appraisal procedure from the project identification, pre-feasibility and feasibility studies, and design, before implementation begins.

This multi-stage process should allow all problems to be ironed out, successive consultants obtaining contracts on the basis of experience and competitive bidding, and checking their predecessors' findings. In practice, the project appraisal procedure is far from perfect. First, badly framed terms of reference can constrain the range of options the consultant is prepared to investigate, perhaps meaning that viable alternative schemes may not be considered. Thus, if asked to investigate the potential for large scale irrigation in a river basin, a consultant will get small thanks for a study investigating the benefits of small-scale alternatives. Since success in bidding for the next job depends on contacts made during this, there are strong reasons for not rocking the boat by presenting unpalatable appraisals. Second, competition has the effect of paring bids for jobs to a minimum, with the result that 'extras' get cut out. Such extras can often mean EIAs, or at least the resources necessary to make them effective. Third, client organizations in the host country can steer the consultant in various ways, for example demanding changes in a report at draft stage, or simply suppressing it (Penning Rowsell 1981). Thus a report suggesting that the anticipated direction of development in a river basin is unwise might prove unacceptable. Some consultants will be unduly swayed by that: 'in some cases, the results of their activities may be so tailored to meet the needs of their client that the distinction between scientific and commercial judgement is blurred' (Chapman 1981, p. 197).

Pressures of this kind distort the technical processes of appraisal. Perhaps for this reason, projects often achieve a self-sustaining life of their own which carries them through the supposed hurdles of appraisal unchanged. Pitt suggests that, once created, plans 'achieve their own reality and ritual inviolability' (Pitt 1976, p. 41). The Volta River Project was first conceived of when Sir Albert Kitson, Director of the Gold Coast Geological Survey, travelled down the Volta (Hart 1980). He saw a possible dam site, and having discovered bauxite the previous year, he conceived the idea of building a dam for hydroelectric power generation to produce aluminium. He made proposals to this effect in the 1920s, but they were not taken up. The idea was brought out again in the 1930s by a South African engineer who formed the African Aluminium Syndicate, and proposed a dam 40 metres high. The idea was pursued in various quarters through the 1940s, with increasing interest from international aluminium interests, before the decision by the colonial government to commission a survey of the basin in the 1950s, and the eventual decision to build the project (under very different terms) in the 1960s (David Hart 1980).

Hill (1978) castigates African states for littering their landscapes with failed agricultural projects, and yet continuing to experiment with basically identical schemes. She argues that 'governments persist in these costly and unremunerative experiments neither because of a commitment either to modernity or productivity nor because of simple ineptitude but because of their quest for control' (Hill 1978, p. 25). She therefore sees the persistence of failure as a function of the attempt by a 'managerial class' to control the means of production, their chosen mechanisms being technical failures (as well as social disasters).

Even without this kind of analysis, it is clear that the interests of the state can be far from those assumed by the economic basis of technical appraisal. Keith Hart points out that a small economic revenue which is predictable may be highly attractive to the state, and even a project with a sharply negative cost/benefit ratio may be attractive if it is 'a rural manifestation of the state's active presence' (Hart 1982, p. 89). Certainly strategic factors can be important influences on development policy, for example in the case of Brazilian settlement in Amazonia, and the construction of the markedly unsuccessful Bura Irrigation Settlement Project in a remote part of northern Kenya in an area troubled in the past by Somali bandits. More prosaically, perhaps, there is a political justification for settlement projects of this kind which is quite separate from the question of their economic performance.

There is also no doubt that corruption is a significant factor in decision-making by Third World bureaucracies. Wellings describes what he calls 'capital leakage' in the public accounts of Lesotho of such a scale that the report of the Auditor General for 1975-8 was suppressed. Wade (1982) has made it clear that corruption in canal irrigation in India is not some kind of curious aberration, but an integral part of the structure of decision-making and performance. The same may well be true in other contexts. Hart comments:

> Today, international construction firms and development bureaucracies vie for shares of what usually turns out to be very lucrative projects for all concerned (including government officials) - lucrative that is for everyone save the farmers whose work on the land can never possibly generate a surplus large enough to pay for all these overheads.
>
> (Hart 1982, p. 94)

Large projects mean large payoffs, and it is clear that in some circumstances this can be an important influence on decision-making.

One example of the complexity and unpredictability of project approval is the Turkwel Gorge Dam in Kenya. Hydrological investigations began in the Turkwel basin in the 1950s, and from the

1960s a number of small irrigation schemes were developed along the river in Turkana (Hogg 1983, 1987a, b). However, it was only in the 1970s that serious attention was paid to the idea of a dam to generate hydroelectricity and an FAO hydrological survey, and subsequent consultancy studies were carried out. The eventual design by SOGREAH involved a dam 300 metres high in the Turkwel Gorge storing 15 billion cubic metres to generate 106 MW of hydroelectricity. In 1986 it was announced that the construction contract had been awarded to a French company at a cost of $205 million (Africa Confidential 1986). The contract was subsequently reported to be worth $270 million, with a supervision contract worth a further $20 million (Africa Confidential 1987).

The letting of the contract and the terms of the loans and credit agreements associated with it have raised considerable controversy both in Kenya and within the donor agency community. The expectation had been that the European Community (EC) would be the lead donor for the Turkwel Project, and discussions took place in 1985 and 1986 both about finance and about environmental impacts. The Turkwel River not only supplies several small irrigation schemes on which a great deal of Kenyan Government and foreign aid money has been spent, but it also feeds a riparian forest which is a vital element in the pastoral ecology of the Turkana. This forest depends on groundwater, and recharge from river flows at flood stages. The dam will reduce the magnitude and frequency of such floods, and as a result forest regeneration may be impaired, degradation of the important grazing resource may take place. Part of the EC deal on the project was an EIA which would assess the significance of these impacts. This would undoubtedly both raise the cost of appraisal and delay the commencement of construction.

However, from 1985 the Kenyan Government had also been negotiating directly with the French Government. In January 1986 a deal was announced covering both design and construction by French consultants and contractors, financed through a mixture of commercial loans and buyers credit from a consortium of twelve banks backed by the French export credit agency Coface and the Banque Française de Commerce Extérieure (Fitzgerald 1986). There was no provision for environmental impact studies. A supposedly confidential memorandum by the EC Delegate in Nairobi, which was widely leaked, suggested that this financial deal was extremely unfavourable to the borrower. Loans appear to be at commercial rates, with a significant proportion repayable in Swiss francs. It was argued that competitive bidding could have reduced the contract price significantly, while a package of grants and soft loans from different aid agencies could have been fitted together which would have been far cheaper. Certainly Turkwel costs compare badly with the $350 million price of the 125 MW Kiambere Gorge Dam, where competitive bidding reduced contract price well below the

World Bank's estimate, and where only half the funding was at commercial rates (Fitzgerald 1986, Africa Confidential 1987). In some ways more serious than the high cost of this dam is the lack of environmental appraisal. Even if other agencies are persuaded to fund an EIA at this stage, all that could be done is to recommend operating procedures for the dam. There seems once again little chance of such a study influencing project design.

Reform of environmental assessment of projects must have several features. Clearly the lack of data needs to be tackled, and this requires long-term monitoring and baseline research. Much of this can be done by national institutions, although pump-priming and support funding may be necessary in the form of foreign aid. A great deal of British aid money still goes on the support of expatriate technical staff. It is probable that Third World countries would be better served by support for indigenous institutions and provision for the sharing of expertise between institutions, particularly internationally within the Third World. There can also obviously be a series of progressive reforms in the way EIAs are fitted into the project planning process: they need to be better-resourced, begun sooner and given time for completion, and to be truly interdisciplinary. They also need to be led by people with appropriate skills. Perhaps this will only be achieved when it becomes routine for the management of appraisal projects to be done by disciplines other than economics and engineering. Another approach is to cease to try to put together multidisciplinary teams by simply aggregating diverse skills, but to seek to train people with multidisciplinary skills and experience.

Above all, perhaps, there is a need for greater openness and for a willingness to learn from mistakes. No-one in development is willing to be proved wrong. Chambers suggests that intelligence is too often defined as 'avoiding being demonstrably wrong about anything' (Chambers 1977, p. 411). Pork-barrel politics is the norm and not the exception, particularly for heads of state, in much of the Third World, and both politicians and bureaucrats share a fear of association with failure. Retrospective studies of environmental impacts are vital (UNESCO 1976, Ashby 1980), not simply as an end in themselves, but as a step towards impact management. EIA procedures should be seen as the start of a process and not the end-point. Few development projects evolve without substantial shifts of design, extent, or speed of implementation; indeed, the failure of appraisal processes generally to cope with the unpredictable elements in development is a major problem.

Barnett (1980) suggests development should not be seen as a kind of black box process, where known inputs create entirely predictable outputs, but as a kind of Pandora's Box. Investment produces change, but it is not possible to predict the extent and direction of that change.

In other words, development is a probabilistic process, full of uncertainties. If development projects are not to fail, they must be allowed to evolve as circumstances change. Environmental appraisals need to evolve with them as they change. Development is a continuing process, and development planning is a form of permanent crisis management. Both development planning and environmental appraisal need to aim for the same flexible and interactive relationship with change.

Aid agencies and environmental policy

The environmental policies of aid agencies have a disproportionate importance in the Third World because much of the Third World is dependent on aid donations to finance major projects. Aid agencies are particularly important sources of funding for the investigative stages of development planning, for example project definition and appraisal. Aid agencies often pick up those aspects of development projects without a direct contribution to economic return, among which the assessment of environmental impacts is notable. The attitude of those First World institutions therefore has a considerable impact on the nature of environmental assessment in development planning. Furthermore, aid is often tied to the involvement of First World consultancy companies, and the aid agencies are thus common routes through which 'experts' are channelled into development project planning.

The performance of aid agencies has been the focus of interest on the part of non-governmental environmental organizations for the past decade (Kennedy 1988, Horberry 1988). The International Institute for Environment and Development (IIED) has been particularly active, conducting research into the environmental policies of both multilateral donors: the major international banks (Stein and Johnson 1979) and bilateral donors (Johnson and Blake 1980). There has also been growing pressure by environmental groups for reform of aid agency environmental policy. This has been particularly strong in the USA, aimed at both USAID itself and the World Bank, whose income is dominated by the size of the American contribution.

The World Bank (or more properly the International Bank for Reconstruction and Development and the International Development Association which are separate arms of the Bretton-Woods family of organizations) is the largest multilateral aid donor. The sheer size of its lending makes its environmental policy of considerable importance, and of consuming interest to environmentalists. Bank lending policy has undergone a number of shifts, the most notable being the move in the 1960s away from exclusive concentration on large scale capital-intensive infrastructural projects to projects designed to aid the poor.

Stein and Johnson suggest that the IIED study of nine multilateral donors was undertaken at a time of 'sharp intellectual transition' for the banks caused by a redefinition of purpose to embrace goals of concessional development assistance and basic human needs (Stein and Johnson 1979, p. 133). The addition of environmental considerations to the cocktail of concerns is therefore problematic. As a result, most institutions lacked clear procedures and criteria for environmental impact assessment. None had developed new forms of accountancy or analysis to bring longer-term environmental effects of development projects into consideration in project appraisal.

The IIED study picks out the World Bank and the Inter-American Development Bank for their greater awareness of environmental problems than others, and their more sophisticated environmental appraisal procedures (Stein and Johnson 1979). The centre of the World Bank's evaluation procedures was the Office of Environmental and Health Affairs (OEHA), but the key to the actual level of consideration of environmental aspects of project development depended chiefly on the attitudes and perception of particular individuals within the normal planning structure: Stein and Johnson argue that 'a wide gap remains between the increasingly alert concern of individuals and the official response of most institutions' (Stein and Johnson 1979, p. 133). This observation is shared by many commentators on the environmental performance of aid agencies. It belies the fact that large bureaucracies such as aid agencies are extremely difficult to reform from within; indeed that there is a certain benefit in having critics on the payroll. The Bank certainly has a somewhat surprising record for employing publicly committed environmentalists like Robert Goodland (see for example Goodland and Irwin 1975), and the economist Herman Daly (Holden 1988). Such mavericks are 'good for the image but mouse-sized in impact' (Watson 1986, p. 275). Watson describes her experiences as an ecologist working for the Bank. She says project staff treated environmentalists 'like scourges', but that nonetheless their ideas were rejected as unrealistic and impractical every time they contained implications for practice: 'by and large we were viewed by project and technical staff as an unessential office which at best put useless icing on the cake and at worst could halt or slow projects' (Watson 1986, p. 269).

Certainly the IIED study reveals the weakness of the OEHA in practice and its failure to have much influence on project design. The OEHA was represented on few World Bank project identification missions, and mission reports suffered from the constraint of 'paying almost unique attention to financial and economic considerations' rarely discussing environmental issues in enough detail (Stein and Johnson 1979 p. 17). The OEHA received the project brief, and thus advanced notice of the general nature of a project. Project preparation is in theory the recipient country's job, but in practice it was often ghosted by the Bank. The adequacy of consideration of environmental matters in the preparation phase depended on the experience of project planners, but it could be cursory, particularly when done by less-aware regional office staff.

In theory, environmental aspects of the project, especially their economic spin-off, were considered in project appraisal. However, the economic data on which the project could be judged at the appraisal stage were both more clear-cut and operationally more workable than environmental data. Few projects have been rejected on environmental grounds alone. The project appraisal document was approved by division chiefs, and the OEHA had opportunity to comment before it went to the Loan Committee of Vice-Presidents of the Bank. Terms are then negotiated with the borrower and the project is given final approval by the Executive Directors. Although, in theory, projects can be sent back for reappraisal, by the approval stage 'many of the design details are immutable' (Stein and Johnson 1979, p. 20). It is effectively impossible to redesign projects, and difficult to build in environmental safeguards. The pressure to keep projects on schedule is intense, and delays 'are most unwelcome' (p. 20).

Clearly, procedures for considering the environmental impacts of World Bank projects within the project cycle were far from ideal at the end of the 1970s. Furthermore, unless environmental issues had been raised in planning, project evaluation missions sent to monitor projects after implementation often failed to identify them, often for the obvious reason that they lacked environmental experts on the team. This was particularly a problem where impacts were subtle, remote from the project or delayed. The IIED Report came up with a series of recommendations, starting with a call for multilateral donors to 'publicly commit themselves, at the highest level' to environmental protection and environmental improvement (Stein and Johnson 1979, p. 137). This commitment should be accompanied by improved procedures and methodologies for environmental appraisal, for evaluation, and the fostering of environmental institutions in developing countries.

In 1984 the World Bank adopted its first official statement on environmental aspects of its work (World Bank 1984c). This took the form of an Operational Manual Statement (OMS). With the exception of a specific mention of environmental audits of completed projects, this effectively set out the status quo on environmental procedures through the project cycle. The Bank at this time treated projects individually, in the context of local setting and 'the ability of the authorities concerned to manage the environment (World Bank 1984c, p. 3). The OMS also drew attention to existing Bank guidelines which 'suggest acceptable ranges to be followed in Bank operations unless the borrowing country's standards are stricter' (World Bank 1984c, p. 3). These principles include for example the provision that the Bank will not finance projects 'that cause severe or irreversible environmental degradation', or which will affect the environment of neighbouring countries, or which would 'significantly modify' Biosphere Reserves, National Parks or other protected areas (World Bank 1984c, p. 4). It was also stated that 'project designers, consulting firms and Bank project officers are expected to provide effective and thorough environmental input into project design construction and operation' (Goodland 1984, p. 13). Environmental work was now the responsibility of the Environment and Science Unit, who primarily worked with regional staff on project appraisal and monitoring (Goodland 1984).

Despite these reasonable-sounding provisions, environmental groups maintained their critical pressure on the Bank (e.g. Goldsmith 1987). Thus in September 1986 a conference in Washington, the Citizens Conference on Tropical Forests, Indigenous Peoples, and the World Bank, voiced numerous severe criticisms of the World Bank's environmental performance, particularly in Amazonia (Horowitz 1987). In 1987 the President of the Bank, Barber Conable, gave a speech to a World Resources Institute dinner announcing a major reorganization. This would involve the creation of an Environment Department and the appointment of forty new staff. In addition, new scientific and technical staff would be appointed in regional offices in Asia, Africa, Latin America and Europe, and there would be money for research on environmental degradation, for desertification control in sub-Saharan Africa, for forestry and for environmental protection in the Mediterranean (Holden 1987). Behind these organizational changes lay some softening of the rigid doctrines of Bank economics, with the appointment to the Latin American office of the zero-growth economist Herman Daly (Holden 1988). New systems of national accounting are being developed to reflect the depletion of non-renewable resources and unsustainable exploitation of renewable resources, and to revise the discount rate which currently biases appraisals in favour of projects with short-term payoffs against longer-term cost/benefit considerations (Holden 1988).

The IIED study of bilateral donors was carried out in the late 1970s, and concerned the Canadian aid agency CIDA, USAID and various other agencies. The agencies vary greatly in size, with USAID by far the largest (a budget of $3.6 billion in 1979). They also vary in the proportion of aid given through multilateral channels and the range of countries supported. Most countries specialize in policy areas where they have particular expertise. For example CIDA specializes in railway and power projects and within agriculture, particularly in wheat schemes. The level of concern for the environment also varies. Johnson and Blake found it relatively good, for example, in the case of Dutch and Swedish aid.

Two key problems were highlighted by the IIED report. The first was the domination of the agencies by economists and planners. They have inadequate training and experience in making multi-sectoral judgements, often making them 'incapable of appreciating the need for adequate management of environmental affairs' (Johnson and Blake 1980, p. 14). The second problem was a shortage of technical staff (related to budget stringency) which led to overwork, ineffectiveness of technical appraisal and lack of opportunity to add skills in environmental appraisal as required. The lack of expertise could be solved by staff training, the recruitment of staff with new skills, and additional help (either in-house or by outside contract) for staff with decison-making responsibilities to present environmental analyses. Beyond this what was needed was better-defined environmental policy, new evaluation procedures ('definite, simple, technically understandable criteria', Johnson and Blake 1980, p. 16) and above all the promotion of indigenous capacity in aid-receiving countries to understand and solve environmental problems (Johnson and Blake 1980).

Resource problems are probably experienced least by USAID, partly because of its size. USAID has a large field staff based in Third World countries. This provision makes it possible to base aid-giving on an effective dialogue between donor and recipient, and creates the possibility of effective consideration of environmental aspects of development. USAID works to a tightly formulated body of regulations, and it reflects the current policy concerns of the US Congress. As a Federal agency, USAID is subject to the provisions of the National Environmental Policy Act (NEPA). Initially USAID established a committee on environment and development, however, it argued that NEPA was primarily intended to relate to domestic developments (Horberry 1988). This view came under pressure from the Council for Environmental Quality and the Center for Law and Social Policy from 1971. Legal action in 1975 made USAID adopt a more positive policy position. In 1978 Congress broadened the Act to embrace the environment and natural resources, and specifically mandated USAID to address deforestation and soil erosion. The extra-territoriality of NEPA was confirmed, and USAID adopted a much wider and more positive environmental policy (Horberry 1988).

The British Overseas Development Administration (ODA) has been the target of criticism on a number of fronts and of calls for reform of aid policy, for example by the Independent Group on British Aid (Elliott 1982, 1984). Critics lament the decline in the amount of British aid (from 0.52 per cent in 1979 to 0.33 per cent in 1987, well below the UN target of 0.7 per cent), and the amount of 'tied' aid and loans under the 'aid trade provision', where the aid is packaged to support a British bid for a contract. In the past ODA has been rich in technical expertise in the Land Resources Development Centre and the Tropical Development and Research Institute. Budgetary cuts have been a problem for the effectiveness of these institutes, and in 1987 it was announced that these would be merged on one site as the Overseas Development Natural Resources Institute, which it can be argued may enhance their performance (Patten 1987). ODA claims that almost a third of its staff are professionally qualified natural resources advisers or scientists (ODA 1987), but, nonetheless, a declining aid budget and the vast range of ex-colonial territories supported means that ODA's technical input is spread thinly on the ground. In 1986 a Natural Resources and Environment Department (NRED) was established to strengthen ODA's consideration of environmental factors under a Chief Natural Resources Adviser, and a single Environment/Research Adviser was appointed (ODA 1987). It is expected that the NRED will develop ODA's policy on the environment, and a procedure has been devised to give early warning of the environmental impacts of projects (ODA 1987).

Critiques of the environmental policy of aid agencies have tended to merge into wider advocacy of sustainable development. The report of the World Commission on Environment and Development (Brundtland 1987) appears to have had an effect on a number of bilateral aid agencies, encouraging the tightening or more explicit presentation of environmental policy. Despite the 'greening' of the rhetoric of development aid, for example the British Minister for Overseas Development declaring himself 'moderately enlightened or be-greened' in 1986 (Patten 1987), there is room for some scepticism about the extent of the transformation of policy. In 1986 the World Conservation Strategy Conference held in Ottawa highlighted a series of policy objectives for aid agencies remarkably similar to those listed by IIED six years before. To realize adequate environmental policies aid agencies should:

- establish adequate professional staff resources to consider environmental issues;
- create suitable guidelines and policy procedures which could embrace environmental factors;
- promote popular participation in planning and implementation of development;
- reform economic assessment procedures to take account of longer-term environmental factors;
- integrate development and environmental objectives.

Clearly a number of aid agencies, multilateral and bilateral, have gone some way down this road, reforming their own procedures and attempting to change practice. It remains an open question, however, as to how much influence aid agencies actually have on the nature and course of development projects. The power of aid donors is often exaggerated, and of course varies a great deal. Mosely (1986) gives an account of the efforts of both the World Bank and USAID to influence domestic economic policy in Kenya. Kenya needed Structural Adjustment Loans because of a balance of payments crisis at the end of the 1970s, and the World Bank tried to make them conditional on adherence by Kenya to the recommendations of the Berg report (World Bank 1981) about the rationalization of industrial protection and the improvement of management in the public sector.

One of the three conditions had been met by the time the second SAL was due in 1982. On this occasion the Bank again asked for an export promotion programme, the liberalization of maize marketing and land reform. The Kenyan Government eventually agreed to terms, which apparently demonstrated the power of the lender. However, in fact the required policy reforms did not materialize, thus demonstrating a corresponding weakness and the lack of sanctions. USAID also tried to persuade the Kenyan government to liberalize maize marketing by breaking the monopoly of the Kenyan Farmers' Association (Cooperative) Ltd, in return for fertilizer. There were strong domestic political reasons why this was an unpalatable route for the Kenyan government to take (not least the power of grain merchants who could get licences). Eventually other factors, including the USA's desire for access to Mombasa as an Indian Ocean harbour, appear to have carried the day. Once again, an attempt to impose conditionality failed.

What this means in practice is that just as the desire of aid agencies to take environmental factors into account in project planning varies, so too does their capacity to do so. The example of the Turkwel Gorge Dam in Kenya discussed earlier is ample demonstration of the failure of an aid donor (in this case the EC) to make environmental appraisal a condition of loan, in the face of eager commercial lenders. There are of course also examples to be found of aid agencies without these constraints failing to carry out environmental impact studies. One example of this, also from Kenya, is the development of projects on the Tana River in Kenya. Hughes (1983) argues that although it was known from an early stage that a series of hydroelectric schemes and an irrigation scheme would have potentially serious environmental effects on downstream floodplain environments, funding agencies (including the World Bank, the EC and ODA) consistently failed to carry out the studies necessary to determine their significance. ODA did insist that a proposed study by UNEP be carried out. This study suggested that the construction of the Masinga Dam could go ahead, but it recommended that a series of further more detailed studies should be a condition of loan, and that an environmental monitoring and control group should be set up. After confrontation between donors and the Kenyan government this idea was scrapped (Hughes 1983).

The other critical issue relating to development in the Tana basin, of the impact of fuelwood demand from the irrigation scheme on riparian woodland, was raised in successive project reports, but the necessary detailed research into impacts was not funded (Hughes 1983). At no stage was a formal EIA carried out on developments in the Tana. Hughes is left wondering if the current concern for EIA on the part of aid donors is a smokescreen, and that in practice 'little pressure is put on governments receiving aid to carry out high-quality environmental studies and monitoring' (Hughes 1983, p. 181).

Clearly, neither a general intention to improve environmental appraisal nor a superficial awareness that environmental impacts may occur are enough to guarantee that aid agencies will take the steps necessary to control the impacts of the developments they fund. Large bureaucracies are inherently conservative, and the 'greening' of development is bizarre theoretically in the context of the established disciplines of development planning, troublesome in terms of policy, and highly inconvenient administratively. Reforming the practice of development bureaucracies has to go a great deal deeper than the superficial transformation of rhetoric and terminology.

8

The Blind and the Dumb: Developers and the Developed

Development planning and indigenous technical knowledge

The consistent failure of government ministry and development agency bureaucracies and their 'expert' employees to take adequate account of the environmental impacts of their actions is matched by their frequent blindness to the needs and capacities of those they develop. It is now commonplace in some quarters to laud the skills and understanding of peasant farmers in the Third World. Such praise forms one of the buttresses of environmentalist critiques of the practice of Third World rural development, and a plank of sustainable development thought in works such as the World Conservation Strategy. At its best the enthusiastic celebration of indigenous skills can be made the basis of alternative strategies of development (e.g. Harrison 1987) .

This new orthodoxy of peasant rationality and skill is of relatively recent date. Formerly, attitudes to peasant production tended to be less generous and less benign, as well as less perceptive. When colonial administrators came to tropical Africa, for example, they mostly failed to see order or skill in rural production systems. Practices such as mixed cropping or intercropping presented an image of confusion and poor husbandry, and the cautious risk-avoidance strategies of peasant farmers were dismissed as the result of a stultified conservatism. There were, however, commentators with greater vision. In West Africa, for example, Jones (1936) celebrated the diversity of native farming. Faulkner and Mackie commented in their book on West African Farming that the farmer would 'generally be found to obtain a maximum of return for a minimum of labour' (Faulkner and Mackie 1933, p. 6). Stamp (1938) argued on the strength of a tour of Nigeria that 'native farmers' had a scheme of farming that could afford almost complete protection against soil erosion and loss of fertility. Elsewhere, investigations of shifting cultivation began to demonstrate the dynamics of the system, and the high degree of environmental knowledge of practitioners (see for example Conklin 1954). De Schlippe, describing the shifting cultivation of the Zande in Sudan, wrote 'the teacher of a culture is its environment, and agriculture is its classroom' (De Schlippe 1956, p. xiv). Allan (1965) wrote a sensitive and remarkably undated account of the diverse skills and practices of 'the African husbandman'.

Ideas of this kind reemerged in the 1970s as part of a liberal and populist reaction against the unsuccessful technological triumphalism of rural development practice, for example in Belshaw's call to 'take indigenous technology seriously' (Belshaw 1974), and in the book *Indigenous Systems of Knowledge and Development* (Brokensha *et al.* 1980). In the West African context of Sierra Leone, Richards (1985, 1986) developed the compelling notion of 'indigenous agricultural revolution', and drew a sharp contrast between the high degree of ecological adaptation in Mende swamp rice production systems and the grim comedy of repeated attempts by the colonial and post-colonial developers to transform them. Like other swamp and floodplain farmers in West Africa, the swamp rice farmers of central Sierra Leone have knowledge of a range of rice varieties with different ecological requirements, and a cultivating environment of considerable variability in small stream-fed inland swamps. Upland and swamp farming are integrated to optimize labour use. Farmers show considerable skill in selecting rice varieties suited to particular flooding conditions, and are also highly innovative. Faced with new seed varieties they will deliberately experiment with them by planting them alongside varieties of known requirements and performance, planting across the downslope catena to assess performance in different conditions, and carrying out input-output trials (Richards 1985, 1986).

The skills and resourcefulness of these farmers of Sierra Leone is shared widely by others elsewhere. Thus, for example, Barker and Spence (1988) describe the 'rich and functional environmental knowledge' of Maroon farmers in Jamaica which underpins their 'food forests'. These are multi-tier crop complexes involving tree crops (such as coconut), shrubs (coffee or cocoa) and smaller plants, sometimes of up to fifty species. There has been similar interest recently in the skills of agricultural management in rainforest areas, both indigenous and of more recent origin (e.g. Denevan *et al.* 1984, Gliessman 1984). The breadth of indigenous skills is well seen in the case of Africa, where the pejorative views of colonial rulers were particularly rife. Skills of a high order exist in both rainfed and wetland farming systems, but they are perhaps most obvious in the many forms of indigenous irrigation that occur in sub-Saharan Africa (Adams and Carter 1987, Adams and Anderson 1988). Irrigation appears to have long been important in the extensive wetlands of West Africa. There are wetland rice cultivation systems on the Senegal and Niger Rivers, in the Delta Intérieure of the Niger in Mali, and on rivers in northern Nigeria. There is also long-established rice cultivation in the Basse Casamance on the Senegalese coast, and similar coastal tide-related cultivation in Guinea and Guinea-Bissau (Linares 1981). In the Volta Delta shallots and other vegetables are grown using soil moisture and some irrigation (Chisholm and Grove 1985).

Inland rice cultivation includes planting on both the rising flood (*crue*) and falling flood (*décrue*). There are similar flood-related practices on the margins of Lake Chad based on 'masakwa' sorghum, apparently of considerable age. Sorghum is also important elsewhere, for example in the Sokoto Valley of northern Nigeria (Adams 1986). In East Africa there are a number of similar examples of river-floodplain cultivation, for example the Tana River floodplain in Kenya and the Tana Delta, the Rufiji in Tanzania and the Okovango Delta in Botswana (Adams and Carter 1987). Many of the residual soil moisture crops cultivated in the wetlands of semi-arid West Africa depend on groundwater lifted by shadoof or simple calabash buckets from shallow unlined wells. In East Africa there are hill furrow irrigation systems of considerable complexity and sophistication in various parts of the Rift Valley, for example Konso in Ethiopia, Marakwet and Pokot in Kenya and Sonjo in Tanzania (Sutton 1984, 1986, Adams and Anderson 1988). In the past there have been similar irrigation systems of large extent, notably at Engaruka in Tanzania (Sutton 1984, 1986). There are analogous irrigation systems in the Pare and Taita Hills, and among the Chagga of Mount Kilimanjaro .

Scientific research has tended to vindicate those who recognized the skill of indigenous cultivators. When agronomists and soil scientists finally carried out scientific research on indigenous farming systems, numerous advantages over monoculture in tropical environments were identified (Igbozurike 1978). Thus Jodha (1980), reporting studies at ICRISAT on traditional dryland farming systems showed that their complexity and diversity enabled farmers to meet multiple objectives simultaneously. Therefore, rather than trying to identify alternatives to traditional polycultural systems, scientists should concentrate on developing more and better options for the farmer, and leaving the choice between them up to him or her. Studies on intercropping of millet and sorghum in dryland savanna conditions at the Institute of Agricultual Research at Samaru in northern Nigeria showed the complex interrelations in phenology, ground cover and nitrogen use, and the positive impacts of intercropping on overall yields (Kassam and Stockinger 1973, Abalu 1977, Andrews 1975). Crops such as millet and sorghum, once dismissed by agricultural extension workers, were shown to be amazingly well-adapted to low rainfall, short growing seasons, and desiccation during growth.

Work elsewhere has revealed the importance of mixed cropping and of community action in pest control (Page and Richards 1977), and the importance of farmers' knowledge of and methods of response to drought (Ogoyuntibo and Richards 1978). By the late 1970s a new orthodoxy was emerging in the African savanna, based on a far more extensive knowledge of production systems, constraints and possibilities (Silberfein 1978, Kowal and Kassam 1978). Interest in indigenous agricultural ecosystems and agroforestry in the humid tropics of Asia and South America is now also well established (Bishop 1978, Denevan *et al.* 1984, Rambo 1982), and it is now common for studies of comparative farming systems (e.g. Turner and Brush 1987) to focus attention on indigenous techniques and their adaptation to environment.

The root of colonial attitudes to indigenous farming practice was a confidence in the superiority of Western civilization, and particularly its science. The cause of the failure of many innovations in development was the inadequacy of that science to comprehend the environment of the tropics. De Schlippe argued that 'if the question is asked why modern civilisation has failed to improve African agriculture, the answer must be that it has proved so far incompatible with the environment of the wet tropics' (De Schlippe 1956, p. xv). The work of ecologists in tropical ecosystems has been described in Chapter 2. The level of understanding of some ecosystems, for example African savanna herbivore systems, is now considerable. What is in general less satisfactory is ecologists' grip of human ecosystems and people-environment relations, for example in the debates over desertification and overgrazing (Chapter 5). The inadequacies of Western science have been matched by an even greater and more serious failure by colonial technicians and administrators to appreciate the social and economic context of farmers' actions. Culwick lamented that in Tanganyika the social sciences lagged behind the physical sciences, so that 'we are today in the rather ridiculous plight of possessing valuable knowledge on the physical side which we cannot apply because we do not know how to' (Culwick 1943, p. 1). Such failures of social and anthropological understanding persist of course today, as Kirby's study of taboos among the Anufo of Ghana shows (Kirby 1987).

Western science has just been part of the wider impact of Western planning. Sir Hesketh Bell wrote of northern Nigeria in 1910: 'it must never be forgotten that we are "protecting" a people in spite of themselves, and that almost every improvement and development initiated by us is absolutely opposed to all their instincts and traditions' (Hill 1977, p. 27). Haugerud comments: 'practitioners of positivist, empiricist science are the new missionaries who would convert developing countries to Western bureaucratic norms and values' (Haugerud 1986, p. 4). The trauma of development lies in the alien nature of those norms and values, inherent in the modernist attributes of Western notions of development. The impact of such imposed ideas extend far beyond agricultural change, just as they are carried over from the colonial period to the work of the modern independent state. Crehan, for example, describes the alien nature of formal schooling in northwest Zambia, in which the curriculum is 'alien not only in its content but in its very form' (Crehan 1985, p. 99).

Population resettlement and the socio-economic costs of development

Development programmes and projects create both winners and losers (Lea and Chaudhri 1983, Hardonjo 1983). Conventional economic accounting attempts to assess the balance of costs and benefits. While this might make sense in terms of national accounting, it disregards the fact that costs are unequally shared. Those who bear the costs may not be those who enjoy the benefits of development. National or regional interests may conflict with the interests of those immediately and adversely affected by the immediate context of development (Cernea 1988). This is clearly seen in the case of population resettlement, particularly that associated with reservoir construction (see for example Brokensha and Scudder 1968, Colson 1971, Scudder 1975, Barrow 1981). Despite what have in some cases been extensive planning exercises, reservoir resettlement has often been far from satisfactory. As Gosling comments in the the context of the Mekong Basin, 'reality is harsher than dreams, and the opportunities for evacuees are more limited in reality than on paper' (Gosling 1979a, p. 119).

Scudder rightly points out that compulsory resettlement is traumatic, causing 'multidimensional stress' (Scudder 1975, p. 455). This stress arises from the way in which people are uprooted from homes and occupations and brought to question their own values and behaviour, and the power of their leaders. Cernea points out that 'in recognition of the hardship and human suffering caused by involuntary resettlement', World Bank policy is to avoid or minimize it, and explore alternative solutions (Cernea 1988, p. 4).

Although they can be relatively short-lived (Barrow 1981), the severe stresses of forced resettlement can be intense (see Lumsden 1975). They are also difficult to generalize about satisfactorily. Colchester (1985) describes the impact of hydroelectric projects in Maharashtra and Madya Pradesh in India on Gond tribal people, discussing in particular the religious importance of certain areas of rainforest. As Sutton (1977) points out, there is little difference between forced resettlement for 'developmental' purposes and anti-guerrilla resettlement such as the *regroupement* policy undertaken by the French army in Algeria between 1954 and 1961. Both are externally imposed and in the end both often require coercion. It is markedly easier to move people to new centres than to keep them there; the act of relocation is the beginning and not the end of the resettlement process.

The key problem in reservoir resettlement is the unequal way in which project costs and benefits are allocated. Gosling (1979a) suggests that resettlement planning should be such that evacuees do not bear disproportionate amounts of the costs of dam projects. The fact that they do so has been the focus of attention by environmentalist critics of dam construction (e.g. Bennet 1978, Goldsmith and Hildyard 1984, Roggeri 1985). Mohun and Sattaur (1987) describe the prospective impacts of the construction of the Bakun Hydroelectric Project in the rainforests of Sarawak. This will involve a dam on the Balui River impounding 4 billion cubic metres of water in a reservoir covering 695 square kilometres. As many as 4,300 Kenyah, Kayan and Kajang people will have to be resettled. The Bakun Dam is one of a series of four dams planned in Sarawak to generate power for Peninsular Malaysia, supplied through an undersea cable.

Lightfoot argues that the literature shows that 'most reservoir resettlements have been badly planned and inadequately financed, and that most evacuees have become at least temporarily and in many cases permanently worse off as a result, both economically and socially' (Lightfoot 1978, p. 63). To an extent the litany of failure may be a function of a lack of published research, and the syndrome that social scientists get more kudos from researching failure than success (Chambers 1983). Nonetheless, there are well-documented case histories, for example from Ghana (Chambers 1970) and Nigeria (Mabogunje 1973).

Gosling (1979a) reports studies of the problems of resettlement following the proposed Pa Mong Dam in the Mekong basin. Planning in the Mekong Basin began in 1957 with the formation of the Committee for the Coordination of the Investigations of the Lower Mekong River (the Mekong Committee) by the UN Economic Commission for Asia and the Far East. A basin plan was completed in 1972, and it triggered concern among environmentalists (Bardach 1973). At 1978 prices the Pa Mong Dam would cost $2 billion to build, and provide 4,800 MW of hydroelectric power and allow 98,000 hectares of irrigation (World Water 1979). The resettlement problems would be immense. The dam would create a reservoir of 3,730 square kilometres in Thailand and Laos, and require resettlement of 0.4 million people. Studies of analogous government settlement projects in Thailand showed settlers remained disadvantaged and poor, unable to regain the income level they had before relocation (Dunning 1979). Similar problems of failure by resettled households to recover previous income levels existed in three small towns in Thailand moved as part of other reservoir resettlement projects (Hafner 1979). This experience tends to support the view that although resettlement planning tends to be highly consumptive of time and resources, it is rarely successful or adequate.

The reasons for the dismal record of reservoir resettlement projects are manifold. The most fundamental is the disciplinary bias of dam designers. Technical disciplines such as engineering, geology and hydrology dominate the dam project planning field, and the appraisal process concentrates on technical problems relevant to these disciplines. The characteristics of the population of the resettlement area, their economy and society, are neither recognized or understood. Field investigations concentrate on the dam site and not the inundation area. Frequently the only research in the reservoir area itself concerns bedrock geology and a perfunctory check on topography to confirm the integrity of the reservoir's proposed storage level. The engineering companies called in at each successive stage of project appraisal lack the skills necessary to comprehend resettlement planning problems.

A second cause of the failure of resettlement projects is the fact that project appraisal rarely allows sufficient time for effective planning to be done. Socio-economic planning is both more difficult and more time-consuming than technical 'experts' assume. It is therefore rarely made part of the technical planning process. It is usually introduced too late to be effective. In the case of the Volta resettlement, for example, although the preparatory Commission had made recommendations about surveys for resettlement, uncertainties over the future of the project meant that nothing was done until the construction contract for the dam was awarded in 1961. The first resettlement staff were appointed in that year, but effective resettlement work did not begin until nine months after the start of dam construction. The stress of resettlement is exacerbated when resettlement is rushed. The human costs of a crash resettlement programme are great, particularly on the elderly and infirm. Furthermore speed brings errors, and problems of food shortage and poor water supplies are often increased (Scudder 1975).

Resettlement planning typically involves a series of tasks, including a population survey and an inventory of property and land within the reservoir area, and surveys to locate possible new settlement sites either near to or further away from home. Only at this stage can the scale of the resettlement problem be assessed. That knowledge is of course a vital element in the assessment of the practicability and acceptability of the project as a whole. Such data on resettlement needs are rarely available in a suitable form at a time to influence decision-making.

A third problem is that of cost. Resettlement is expensive, even if done badly, and it is usually under-resourced. This is partly because costs can only be calculated once preliminary surveys have been carried out. But these surveys only begin after the decision to go ahead with the project has been made. Resettlement costs are therefore seen as somehow an added extra, additional to the costs of construction. It is as if they were in some way optional. An interesting example of this is the development of the Volta scheme in Ghana. The Akosombo Dam was begun in 1961, but was preceded by an extensive study published in 1956 by a government Preparatory Commission. This recommended that extensive further studies be carried out into resettlement problems, but suggested that self-help resettlement with cash compensation might be the best solution. Partly because of its concern with resettlement and the resulting costs, the work of the Preparatory Commission made the economic prospects of the dam seem sufficiently unattractive that work did not proceed. Within a few years, however, in the aftermath of an agreement between the USSR and Egypt over the Aswan High Dam, and in an independent Ghana, the scheme took on a new aspect. Revisions of the Preparatory Commission's figures were made which excluded a number of costs, and work began (Hart 1980, Lanning and Mueller 1979).

A fourth problem is that, in the final analysis, resettlement is a political and not a planning problem. Its sensitivity may make it unattractive to able civil servants and lead to it being both farmed out to consultants and shielded from the attention of researchers (Barrow 1981). The success of the whole exercise depends on the existence of an adequate legal regime, for example one laying down principles of compensation. Obtaining such a regime can take time, yet without it the resettlement project will lack any legitimacy. Costs will depend on the nature of the legal regime, and the extent of financial compensation, compensation in kind (for example replacement housing, water supplies or farmland) and logistic support that is offered.

Lack of public particiption in planning, and lack of advanced warning of project development in the reservoir area, combined with inadequate (perceived to be unfair) provisions for compensation are major sources of grievance with resettlement projects. Their lack of 'effectiveness' in planning terms is matched by their lack of legitimacy. Gonzales (1972) describes the complex social context in which a hydroelectric dam in the Dominican Republic was built. He identifies diverse interests in the dam, and points out that while it will not fulfil the dreams and expectations of the peasants (since the benefits will accrue to large landowners and urban dwellers), they are firmly in favour of it. This paper is a salutary reminder of the unsatisfactory nature of simplistic comments about the impacts of project development.

Although in some cases even those who will be adversely affected by resettlement show initial enthusiasm, opposition is more common.

Howarth (1961) provides a rather bucolic picture of local opposition to resettlement in the case of Kariba. A less cosy example is provided by the Bakolori reservoir in northern Nigeria, built to supply water to an irrigation project. The dam was completed in 1978, flooding the homes of 12,000 people. There was no attempt to involve those affected by the reservoir in planning, and no specific provision at design stage to set out a specification for resettlement. Survey in the reservoir area was left to the river basin authority, who lacked technical skills. Most of it was in fact done by teams of students, and surveys were only narrowly completed in advance of the waters of the filling reservoir (Adams 1988a). Farmers complained about the inaccuracies and haste of the survey and the inadequacy of the compensation that was paid.

Disputes about compensation persisted for several years. Evacuees were settled in a series of large settlements on barren hilltops remote from the river valley (Wallace 1980), without cultivable land. They abandoned the resettlement villages in considerable numbers. Eventually water was lifted from the dam at great expense in an attempt to irrigate land adjacent to the largest village of New Maradun, but in fact by that time (1980) the complaints of evacuees had become inextricably linked with the separate grievances of those in the irrigation area served by the dam (Adams 1988a). The resulting blockades of the project area by farmers, and the violent repressive action by Federal police, are described later in this chapter.

The lack of consultation at Bakolori may be contrasted with the method adopted in the Volta resettlement to select new village sites (Chambers 1970). The Volta resettlement involved some 80,000 people in 739 villages, about one per cent of the population of Ghana. As part of an extensive social survey, enumerators recorded the positive and negative preferences of each village about relocation. Seventy-two sites were identified in this way; of these twenty-seven were rejected by the Volta Resettlement Authority on technical grounds, although some were accepted after further discussion. Fifty-two village sites were eventually agreed.

Very often resettlement projects begin with high hopes of re-establishing evacuees in conditions no worse than those they have left. Thus the Resettlement Working Party convened by the Volta River Authority aimed to use resettlement to 'enhance the social, cultural and physical conditions of the people' (Chambers 1970). Such high hopes are rarely realized. In the Volta case they led to enthusiasm for mechanized farming, which proved one of the least successful features of the resettlement. It was estimated that 42,000 hectares would be needed to support evacuees. However, land clearance was slow. By 1967 only 2,500 hectares were being cultivated, over half of them manually. In 1968 only 52 per cent of the adult males resettled could farm at all, and food relief had been necessary for three years (Hart 1980).

Failure to specify the basis for resettlement and the lack of data on the affected population leads to gross underestimation of the costs of resettlement. Inadequate budgetary provision is made, and the lion's share of this is eaten up by the survey and planning elements of the resettlement process. This was the experience in another Nigerian scheme, the Dadin Kowa Dam in Bauchi State. Here 26,000 people were to be resettled. No planning for resettlement was done at the stage when the dam was being appraised and designed. Construction began in 1980. In that year a preliminary resettlement 'Master Plan' was completed by one consultant, who was then replaced by another who began the detailed planning. Within three months it was clear that the cost of compensation at the rates set by the Federal Government would be about 60 million Naira (then about $50 million), more than an order of magnitude greater than the total resettlement budget. On top of that there would be a great number of substantial additional costs of infrastructural development (for example roads and water supplies in new villages). Even then, all that would have been achieved would be relocation, since such a resettlement was far from recreating a viable economy or acceptable living conditions.

Obviously resettlement was a very significant element in total project costs; if tackled seriously it would have altered the cost/benefit calculations of the dam irrevocably. In the event, attempts to seek additional funds from the Federal Government were overtaken by the fall in world oil prices and severe financial stringency at Federal level. The resettlement planning became a paper exercise, and was not implemented. The whole affair was greatly complicated by the fact that two separate states and two river basin development authorities were involved in overseeing the planning, and furthermore that the dam itself was embarked upon before basin-wide planning was begun. When that was done, it was concluded that it was not an appropriate place to make investment on this scale; such a dam would be better higher in the basin to achieve flood control (Adams 1985b).

The impact of changing planning structures and approaches is better documented in the case of the Kainji Dam in Nigeria. This was completed in 1968, creating a lake of 1,200 square kilometres, flooding 203 villages containing 44,000 people, the towns of Bussa and Yelwa, and 15,000 hectares of farmland (Adeniyi 1973). Resettlement planning began before independence in Nigeria, done by the Ministry of Economic Planning of the Northern Region. Subsequently it was taken over by a federal Niger Dams Resettlement Authority and after 1966 by the Military Governor of Kaduna Region. Initially, evacuees were to be compensated in cash, and surveys began in 1962. By the end of 1963, 2,338 people in eighteen villages had been moved. However, resettlement was slow and was deemed to be inefficient. In November 1964 the policy on housing was reversed, and the Authority began to build houses for evacuees using a design by British architects using sandcrete block walls and ferro-cement roofs (Atkinson 1973). Cash was paid for farmland and trees of economic importance, and agricultural advice was offered to evacuees.

The resettlement at Kainji appears to have been moderately successful, the main immediate focus of complaints being poor village locations and the design of the new houses, particularly the layout of rooms, their thermal properties and leaking roofs. Indeed, the Kainji resettlement experience remains one of the relatively rare success stories of paternalistic central planning of population resettlement. It shares this reputation with the Volta Project. In neither case has there been a great deal of long-term monitoring, and most evidence comes from surveys soon after resettlement. The Volta resettlement was based on 'core' houses. Studies in one town, New Mpamu, in 1965 recorded that 50 per cent of settlers were enthusiastic about their new houses (Tamakloe 1968), but nonetheless social problems were reported (drunkenness and fighting for example); in other places (in a later study) such phenomena were explicitly linked to problems of overcrowding and lack of privacy (Lumsden 1975). Perhaps the best measure of the 'success' of these resettlement projects is the loyalty of those resettled. It is eloquent testimony that of the 67,500 evacuees on the Volta project only 25,900 were still present in resettlement villages in 1968. Other evacuees had moved elsewhere, while outsiders, including fishermen displaced by declining catches downstream, had arrived. Other resettlement schemes exhibit similarly high turnover rates. Evacuees from the Aswan Dam were resettled on the New Halfa Agricultural Scheme (originally called the Khashm el Girba Scheme in Sudan). Of 300,000 people in the area in 1977, about 24,000 had left by 1980. Both the productivity and cultivated area under irrigation began to fall due to a combination of technical and management problems (Khogali 1982).

There are clearly quite specific skills associated with population resettlement planning (Butcher 1971). It has been argued above that these are often lacked by those central to the dam project planning process, but nonetheless the nature and extent of the problem is now widely recognized. Guidelines are now available outlining not only the essential elements of a resettlement project, but discussing their integration into the operational procedures of the project development cycle (Cernea 1988). Goodland (1978) outlines the measures necessary in the fields of social and cultural ecology with respect to the Tucurui Dam in the Tocantins River basin in Brazil. He urges that mitigation of the project's impact on indigenous Amerindian people should be allocated time and resources commensurate with its importance, and that the associated costs should be considered an integral part of the costs of the whole project.

Lightfoot points out that planners assume that their expertise allows them to 'understand and manage the interests of the farmers better than the farmers do for themselves' (Lightfoot 1979, p. 30). As a result resettlement planners tend to neglect consultation and assistance with self-help projects and favour instead direct intervention in the resettlement process. This is despite the fact that among evacuees from the Nam Pong Project (5,012 households resettled in 1964), those who had been resettled in planned schemes were worse off than those who had resettled themselves (Lightfoot 1978, 1979, 1981). In part this may be because it is the least able and experienced, for example the old and unfit, who have to rely on such schemes (Dunning 1979), but nonetheless, the inability of the resettlement planning process to create adequate replacement livelihoods remains a key failing.

Both the Kainji and Volta resettlements began with self-managed resettlement based on cash compensation and abandoned it. In the Kainji case, for example, self-managed settlement was too slow, and villages were sometimes located in sites unsuitable for water supply or other infrastructure. Perhaps more fundamentally, evacuees were using compensation money for purposes other than house-building. Presumably this reflected either the existence of economic opportunities other than reestablishment of a farming household and land clearance, or (more probably) hidden costs facing evacuees. However, it was interpreted as a shortcoming of the laissez-faire approach, and the centralized planning input was strengthened. In most cases, resettlement planning is based on this kind of top-down centre-outwards approach to planning. It forms part of a development process that is imposed from outside, and meets an agenda which may be little influenced by local experience of past or present.

Resistance to development

The costs of development projects can clearly be substantial, and borne most unequally. While economists (and aid donors) worry most about those which have the poorest economic records, even economic successes can involve substantial costs to an element of the population. In some circumstances, these costs can generate opposition. Recent work in sub-Saharan Africa suggests that conflict between developers and developed has been a central feature of the transformation of rural economies by colonial and post-colonial states (Crummey 1986). The social and cultural impacts of development projects has been discussed above. The reality of state coercion of rural producers may be less immediately apparent, but it is in fact widely commented upon (Hill 1978, Williams 1976, 1981). Development may hold little comfort for the rural poor.

There is considerable debate, particularly in Africa, about the nature of peasantries and the political economy of development (e.g. Bernstein 1979, Smith and Welch 1978, Klein 1980). Hyden has argued that the Tanzanian state has failed to capture the peasantry, that 'the capitalist mode of production is by no means so powerful that peasant power is irrelevant for development issues at large' (Hyden 1980, p. 37). He sees an active peasant mode of production that has conferred political power on peasant producers. Others dispute this view, urging the power of capitalism. Thus Good (1986) argues that the Zambian state did capture the peasantry in the 1920s, and made them bear the burden of capitalist development. Many Third World governments have shared the view that peasants 'are expected to contribute to development by providing the resources for others to develop the urban industrial economy' (Williams 1976, p. 131). With an agricultural policy aimed at taxing producers to pay for urban infrastructure, and price controls to keep urban food prices down, the construction of projects such as hydro-electric dams in rural areas may well make the state and its 'development' a highly unattractive actor on the rural scene.

Peasant farmers rejected colonial 'development', as they reject development today, for many reasons. One is its alien nature, and the drastic cultural changes it demands. Another is simply its technical blindness and incompetence. As Baldwin (1957) commented, when settlers were finally sought on the Niger Agricultural Project at Mokwa in Nigeria, after several years of unsuccessful mechanized production of groundnuts, they thought not that they were being offered 'development', but that they were being called to the rescue. Their response was not enthusiastic. The inadequacy of such blundering projects in post-war Africa, most notoriously the Groundnut Scheme at Kongwa in Tanganyika, make the reluctance of peasants to become involved seem entirely understandable. However, even though contemporary commentators like Faulkner and Mackie pointed out that 'the prevalent idea that the native farmer is excessively conservative is largely due to the mistakes of Europeans in the past' (Faulkner and Mackie 1933, p,7), the myth of the conservative peasant ran deep in the colonial mind. The failure of development projects was blamed on this non-participation, the 'intransigence and primitiveness of peasants' (Hill 1978). Williams (1976, 1981) discusses what developers saw - and still see - as 'the peasant problem', and argues that for developers this 'recalcitrance of peasants to outsiders' conceptions of progress and the peasants' place in it defines the peasants as backward and delineates the peasant problem' (Williams 1981, p. 132).

There are few studies of development written from the perspective of the developed, giving the view 'from below'. Academics, particularly anthropologists, sometimes attempt to speak for those without voices in the remote literature of development, but most researchers and almost all development professionals are condemned to some form of 'rural development tourism' (Chambers 1983). They turn up in a village on a tight timetable with a tight budget, seeking something. Whether that something is information, agreement or votes, village people might well regard with considerable caution the stranger in a jeep with city clothes, a bagfull of papers and a clipboard. Interestingly, it was a young researcher to whom Adrian Adams (1979) wrote to explain the experience of the village of Jamaane on the River Senegal of Europeans. He had come from the development agency (the Organisation pour la Mise en Valeur du Fleuve Sénégal, OMVS) filled with their presumptions and philosophy. He was but one in a long chain of visitors, mostly white, mostly staying only short periods, who transformed *un développement paysan* into *un développement administratif*. A peasant-run project based on communal work groups was gradually taken over by the Société d'État with the introduction of mechanized irrigation exclusively for rice, funded by USAID.

To the technician through whom the alienation of indigenous change began, 'develop' was 'an intransitive verb, and "development" was one and indivisible' (Adams 1979, p. 473). The changes he and his successors introduced were 'presented as neutral steps, objectively required on technical grounds as part of the development process'. The technician was an 'expert', someone with 'a halo of impartial prestige' lent by his skills, able to neutralize conflicts and package political issues as technical ones. Such an expert embodies 'modernity, progress, efficiency', and as Adams points out, 'no-one may be an expert in his own country: it's an expatriates' role'. Adams argues that these development visitors failed to acknowledge that Jamaane village existed, that there were live people there ('Ils ne savent pas qu'il y a des gens ici vivants'). No-one, development expert or researcher, wanted to know the truth (Adams 1979, pp. 473-4).

Reluctant participation in the projects of the state grades into outright opposition. As such it forms part of wider peasant resistance, and sometimes revolution. Bates, for example, describes the politicization of the African peasantry, arguing that 'the commercialisation of agriculture was intimately linked with the rise of political protest in the rural areas' (Bates 1983, p. 104). Migdal (1976) reviews the way Third World peasant communities move towards an outward-orientation, involving for example wage labour and involvement in a market economy. He argues that peasant revolution must be seen as part of this reorientation. Hildermier (1979) distinguishes between agrarian revolts, often short-lived and violent, and longer-term and more peaceful agrarian movements. Certainly rural opposition to state intervention is very often spontaneous and localised. In the context of the Transkei, Beinart and Bundy see peasant movements 'limited in aims and achievements, and deficient in organisation and execution' (Beinart and Bundy 1980).

Opposition can also fall very far short of revolution, indeed it can be both peaceful and silent. Spittler (1979) discusses what he calls the defensive strategies of peasants faced with what they see as the unreasonable demands of the state. One such defensive strategy is evasion, for example hiding. In the colonial period lookouts outside villages reported the arrival of soldiers and policemen, and people hid in the bush to evade tax-collection, requisitioning or labour demands, and hid cattle and children during censuses. Another strategy is to ignore orders ('silent disobedience', Spittler 1979, p. 31), for example groundnut smuggling or failure to vote; another to misuse material or money or livestock given for a specific purpose; another to end difficult interviews by agreeing with everything. The state in turn responded with a paradoxical mix of 'laissez-faire and force' (Spittler, 1979, p. 33), often alternated, for example rounding people up who refused to participate in a development project, but eventually dropping it.

There is clearly a complex relationship between resistance (typified by 'silence and stealth', Crummey 1986, p. 10), and protests which are more vocal or active. Rudé writes of the land occupancy protests which followed the alienation of common lands in Latin America by *haciendados* in the 1920s. Initially these were 'more or less spontaneous and localised affairs, involving peaceable squatting in the first place, but almost inevitably developing later, as the troops moved in, into violent confrontations, at great cost in peasant lives' (Rudé, 1980, p. 69). Crummey refers to this as the waking of 'the other beast' of state violence (Crummey 1986, p. 21). Rural opposition to state action is important in the dynamics of rural change in many Third World countries. They occur in particular in response to the expanding centralized bureaucracy of development, to the central role of 'outsiders' and 'experts' in the development process and to the impacts of transnational agribusiness enterprise. They are fundamental to any understanding of the form and impact of government-induced rural change in specific locations at specific times.

There are in the Third World numerous examples of failure in development, and governments and international agencies continue to implement schemes 'little better planned than their more spectacularly misbegotten predecessors' (Hill 1978, p. 25). One example of alleged failure is the social and economic impact of large-scale irrigation in northern Nigeria in the 1970s (Wallace 1980, 1981, Oculi 1981, Palmer Jones 1984, Andrae and Beckman 1985, Beckman 1986). Although economic and technical aspects of these developments have been criticized, it is the social impacts of project development, and more particularly the political responses to those impacts, which are of greatest interest.

The Bakolori Project in northern Nigeria has already been mentioned in the context of the impacts of dam construction on downstream agriculture (Chapter 6), and the problems of resettlement of those flooded by the reservoir (this chapter). The scheme consists of a dam on the River Sokoto to store water to supply an irrigation scheme of 30,000 hectares (gross area). The scheme formed part of a basin-wide study carried out by FAO in the 1960s. This focused on the need for flood control in the lower basin through dams such as that at Bakolori in the headwaters. However, the scheme was enlarged by Italian designers in the 1970s. Land clearance on the area to be irrigated took place on land already farmed with rainfed crops. Farmers complained about the land expropriation and reallocation process, about planting bans and damage to growing crops and above all about compensation for economically important trees and other improvements. From 1978 to 1980 there were repeated blockades of parts of the project area by aggrieved farmers (Adams 1988b). These culminated in a violent police action in 1980 in which a substantial number of people died. While this did not solve the grievances, it effectively ended overt public protest about the development. Beckman says of these events: 'the Bakolori peasant rising and the extreme violence by which it was repressed will continue to cast long shadows over the rural scene in Nigeria, providing inspiration to some and warnings to others' (Beckman 1986, p. 154).

Studies of Bakolori that embrace this protest have placed it firmly in the context of national political economy, and particularly the concentration and centralization of state power over the rural sector in the 1970s. Thus Oculi (1981) stresses the powerful position of the banks and transnational corporations at the time of project appraisal at Bakolori (which came before the oil boom), and the undermining of local planning through the cooption of local state officials (in the Federal River Basin Authority) via 'bribery or ideological collaboration' (Oculi 1981, p. 204). Beckman (1986) develops a similar argument, focusing in the case of Bakolori on attempts by the state to control land and labour, and the common concerns of international capital and the Nigerian elite. Andrae and Beckman (1985) focus on the 'wheat trap', placing the failure to develop large-scale irrigation for foodgrain production in the context of increasing dependence on imported wheat, and the power of milling and baking interests.

The nature of the planning process at Bakolori is also extremely important in understanding the nature of the protests (Adams 1988a). Policy-making is not mechanistic or simple, rather 'the whole life of policy is a chaos of purposes and accidents. It is not at all a matter of the rational implementation of so-called decisions through selected strategies' (Clay and Schaffer, 1985: 192). The question of what actually happens in project planning is seriously under-researched, and ignorance is compounded by the lack of effective communication between academic workers on development and practitioners. An understanding of the planning process is important, as is reform (Clay and Schaffer 1984).

Conservation as development

Sustainable development often starts from a search for ways to redirect development that are more benign for nature conservation. Conservation organizations use ideas of sustainable development to claim to represent the true interests of rural people, who are themselves often the victims of development. Sustainable development is therefore (almost by definition) that which will appeal to local people. However, conservation activities by themselves may be much less attractive. Thus attempts to stop hunting and tackle 'the menace of shifting cultivation' in forest reserves, key elements in a conservation policy for a country like Nigeria for example (Anadu 1987, p. 249), must be made against intense local opposition. As promoted by conservation interests, sustainable development is often seen as a means of winning over local opposition to conservation objectives, the carrot which balances the more conventional sticks such as anti-poaching and land use control measures.

To the rural producer, conservation can readily seem to be yet another form of development. Its priorities are set from outside his community, in the priorities of national conservation agencies or by international organizations. Work is often funded from overseas by specialist conservation aid agencies who employ their own visiting experts. They carry their own assumptions, stay short periods, and suffer from the same biases (seasonal, urban, tarmac) as their economic and engineering counterparts. Repeatedly, such expert missions identify the current actions of local people as a threat to the survival of some feature of conservation interest. The policy proposals of these development-conservationists may be as alien to the conceptions of local people as any proposed by other developers, and as adverse to their interests. They may as a result be highly unpopular.

There are many examples which could be chosen to demonstrate the destructive nature of conservation as development. Most stem, as do their agricultural counterparts, from the strength and impact of the ideologies of conservation or development. The curious psychology and sociology of hunting in colonial Africa is described by MacKenzie (1987). The rise of the hunt as a ritualistic pastime for whites was matched by the dismissal of hunting by Africans as poaching, and its prevention. When ecology developed as a tool of land use decision-making, it was marshalled in the service of proving the hypothesis created by this ideology: that African hunting damaged wildlife populations. Collett (1987) argues that there has been a continuity in government attitudes to the Masai based on this supposition. He finds no evidence for this in the archaeological record. Early colonial administrators developed the view that the Masai were 'predators terrorising neighbouring groups', accumulating stock and refusing to trade (Collett 1987, p. 144). From the viewpoint of the Masai, pastoral policy, particularly sedentarization, and conservation policy are merely alternative forms of intervention by the state.

The role of conservation as a form of development, the more or less unwelcome work of state bureaucracy, is particularly important in the context of the designation of land within protected areas. Collett argues that the Masai opposed state alienation of land both for white farms on the Laikipia Plateau or for wildlife conservation. Similarly, Gordon (1985) describes the impact of nature conservation on the freedom, culture and access to resources enjoyed by the Bushmen of Namibia. Game reserves were first declared in Namibia in 1907, and involved bans on the hunting of giraffe, buffalo, and female eland and kudu. Various game reserves were declared, and in some (notably what became the Etosha Game Park) Bushmen were tolerated, and used as trackers and piece labourers. Elsewhere, they were moved out.

The Gemsbok National Park was declared, in 1931, on land occupied by Bushmen. The National Parks Board banned hunting, and prosecuted 'poachers'. In 1941 land was set aside for Bushmen to hunt adjacent to the park, but there and elsewhere cultural contact and economic integration brought cultural change and attempts to move Bushmen who were no longer 'true' Bushmen living in a traditional manner, out of the parks and reserves. In 1955 the Department of Nature Conservation was established, and in the 1960s white tourism became important. In 1980 the Department began to open up new areas for tourism, including a 562,000 hectare Game Park in Bushmanland, in the Kalahari on the Botswana border. Gordon argues that this park makes sense only by making the Bushmen themselves the subject of tourists' quests, 'the objects of the leisure rituals of the affluent whites' (Gordon 1985, p. 40). The people of Nyae-Nyae will be able to hunt 'traditionally', but not develop their economy to embrace subsistence pastoralism. Gordon titles his piece 'conserving the Bushmen to extinction' (p. 28).

Turton (1987) discusses the relation between the impact of state conservation on the Mursi of the Omo Valley in Ethiopia to the very different cultural values of the Wildlife Conservation Department and the Mursi themselves. This area is perceived by conservationists, quite incorrectly, as 'wilderness'. In fact it is an anthropogenic ecosystem created by the Mursi. Their economy is based on cattle herding, dry season cultivation and flood-retreat farming along the Omo River. All three have to be combined to achieve subsistence. The Mursi also hunt in the hungry season before harvest, and trade ivory, leopard skins and other products.

The Omo National Park was established in 1966 West of the River Omo, and a second park to the East, the Mago National Park, was planned in the 1970s by a Japanese team. Over this period Mursi have been driven south into the Mago area by drought, and some uses have intensified. Despite the parks, the area has remained remote, with few visitors, and the impact of conservation authorities on the Mursi have so far been fairly minor, with some restriction on farming and extra risks in hunting. Turton argues that the National Parks have made the Mursi marginally vulnerable to drought and famine, but points out that this was simply because of the poor grip of the conservation authorities on the area: 'if the integrity of the Omo and Mago Park boundaries had been successfully maintained against human occupation and use, there would simply have been no Mursi economy left' (Turton 1987, p. 178). The 1978 report saw the Mursi as a threat to conservation, although without defining in what way, and proposed resettlement. Turton ridicules the policy of exclusive conservation, and draws a sharp contrast between attitudes to the conservation of wildlife and those towards the survival of the Mursi themselves.

Of course, protected area policy has undergone many changes within recent decades (McNeely and Miller 1984, Nelson 1987), some of the them driven by the adoption by conservationists of ideas about sustainable development. Thus McNeely argues that while National Parks are as important as ever, and 'as carefully protected as ever', they must be supplemented by other kinds of protected areas 'to meet the social and economic development needs of modern society' (McNeely 1984, p. 1). In fact, the IUCN Commission on National Parks and Protected Areas recognizes ten categories of protected area, ranging from the 'strict nature reserve' through the National Park or National Monument to 'resource reserves' and 'natural biotic area or anthropological reserve'. This last category would include areas where people obtain their livelihoods 'by means that do not involve extensive cultivation or other major modifications of the vegetation or animal life' (IUCN Commission 1984, p. 51). This approach was also integral to the Biosphere Reserves introduced by the Man and the Biosphere (MAB) Programme in the 1970s (see Chapter 2).

Biosphere Reserves were to be located in representative areas, and of large enough size 'to accommodate different uses without conflict', including ecological research (Batisse 1982). As such they represent important examples of the attempt to integrate conservation and human use. However, in the Third World, practice has in some cases diverged from principle. Schoepf (1984) reviews the impact of the (MAB) Programme in the Lufira Valley in Zaire. The Zaire MAB Project was established in 1979, and placed in 1982 under the Department of Agriculture and Rural Development. There are three regional ecosystem management programmes in different ecological zones.

The Lufira Valley MAB Reserve covers 50,000 hectares in savanna Miombo woodland in southeastern Shaba. The Lemba people used to integrate shifting cultivation with more intensive practices, including mounding and ridging using mulch, and hand irrigation of streamside gardens. During the colonial period labour emigration was significant, and villages and fields were relocated along roads 'to facilitate administrative surveillance' (Schoepf 1984, p. 272). The Lufira reserve was selected by an expatriate ecologist, and by 1981 had yet to be visited by the Chairman of the committee at Lubumbashi designated to oversee it. Research funded by the Belgian government has concentrated on botany, and is not informed by understanding of the ecology of indigenous production, or the lack of success of past attempts to intensify agriculture. Schoepf argues that ecological researchers repeat colonial misunderstandings about 'irrational peasant traditionalism' (Schoepf 1984, p. 275).

MAB reserves are zoned, with a core where all productive activities are prohibited, surrounded by a buffer zone where existing activities may continue under MAB supervision. However, innovations causing environmental change are banned. Schoepf argues that MAB has become caught up in efforts by the state to increase agricultural production through obligatory cultivation of cassava and maize, and that this would be enforced in the experimental zone. MAB officials adjusted the reserve boundary to exclude charcoal cutting areas of parastatal and private firms, and certain large farms, stating that the central zone was uninhabited. The chiefs of the Upper Lufira Valley complained in 1980 about the establishment of the reserve, and listed 2,000 people who lived or cultivated there. The chiefs were dismissed as uncooperative, obstructionist and anti-state. Schoepf argues that the MAB project has been effectively hijacked by the entrenched class interests which control bureaucratic decision-making in Zaire. He predicts that it will increase inequality in the Lufira valley and promote rural emigration.

The most often cited example of the integration of conservation and development based on conservation areas is in the Amboseli National Park in Kenya. Amboseli lies in land grazed by the Masai in the dry season, with cattle comprising up to 60 per cent of large animal biomass at this season (Lindsay 1987). A massive game reserve was established in southern Kenya in 1899, and the Amboseli National Reserve was created in 1952 as one of a series of smaller reserves within this area. Hunting was banned, but Masai grazing continued. In

1961 the District Council assumed control of the area, but although they received some entrance fee revenue and hunting licence fees, conflict developed because of Masai suspicions that further grazing rights would be lost. To the Masai of Amboseli the District Council was 'only another level of bureaucratic authority' (Lindsay 1987, p. 153). Large game animals began to be killed. The Masai demanded formal ownership of the area, the conservationists demanded a National Park.

In 1974 a 488 square kilometre National Park was declared, subsequently reduced to 390 square kilometres. The area was made part of a complex agreement whereby Masai gave up the right to graze within the park in return for joint ownership of surrounding bushlands in group ranches. These received water supplies, compensation for lost production through wildlife grazing, the development of lodges and wildlife viewing circuits and developments such as a school and a dispensary. The District Council retained control of lodges on 160 hectares in the heart of the Park. Initial reviews suggested that the Amboseli programme was a great success (Western 1982). However, Lindsay (1987) describes a series of problems. Although a school and cattle dip have been built, and wood and gravel have been sold to the Park, and there have been campsite fees, the borehole and pipeline system has been poor and the financial benefit to the Masai has been small. The wildlife utilization fees were paid irregularly after 1981, no new viewing circuits or lodges were created, and park entrance fees were retained by the Treasury. Tourist numbers levelled off in Kenya in the 1970s, and Masai incomes did not rise as predicted. Killings of rhinoceros and elephant began again in the early 1980s, and poaching increased. Lindsay argues that the Amboseli Park Plan, like previous programmes, 'failed to provide the Masai community with continuous appreciable benefit in return for compromises in their use of land' (Lindsay 1987, p. 161). Income to the Masai was too little and too unpredictable, and continuing cultural, social and economic change among the Masai undermined static assumptions about the long-term acceptability and sustainability of the group ranching system.

Clearly conservation initiatives, like development projects, have to be implemented in a real world subject to manifold political pressures. In the past many conservation projects have been insensitive to the needs and aspirations of local communities, and ignorant of practical local politics. However, this problem is precisely that which sustainable development thinking promises to address. Sustainable development claims to avoid adverse environmental impacts, destructive socio-economic impacts, and the systematic land degradation created by poverty. To what extent does sustainable development provide a means of integrating conservation and development objectives in conservation areas?

Largely because of concerns such as this, conservation organizations also have begun to develop policies for tribal groups and others (McNeely and Pitt 1987). This has been part of a wider concern about the specific impacts of development on tribal people, led by groups such as Survival International (Survival International 1985). The reality of development for tribal groups can be grim, for example in areas such as Amazonia (Vickers 1984, Arvelo-Jiménez 1984). Budowski (1982) discusses the impacts of various different forest management practices in Central America and the Carribbean on local people. Both conservation and logging are often viewed as infringements on existing rights. Murdia (1982) describes the impact of forestry policy in India on tribals. In 1952 the rights granted to forest-dwellers in 1894 became concessionary. New policy objectives for the forests were first to meet national goals of industry or defence, second 'to wean the tribals away, by persuasion, from shifting cultivation' (Murdia 1984, p. 34), third to control grazing, fourth to promote welfare and fifth to increase the effectiveness of forest management through laws and law enforcement. The loss of livelihood in farming and the loss of forest resources (fuelwood, building materials, fodder) was not compensated for even in narrowly economic terms by casual labour opportunities.

In 1982 the World Bank issued a paper on *Tribal Peoples and Economic Development*. This stated that their policy was not to develop areas presently occupied by tribal people, but that if this was unavoidable their policy was to ensure that 'best efforts have been made to obtain the voluntary, full, and conscionable agreement (i.e. under prevailing circumstances and customary laws)' of the tribal people or that of their advocates, and that project design and implementation 'are appropriate to meet the special needs and wishes of such peoples' (World Bank 1984a, p. 1). While many observers remain unconvinced by rhetoric of this kind, Maybury-Lewis argues that it should not be assumed that Indian societies are doomed just because of the strength of the ideologies behind the second conquest. He argues that 'we have to come up with strategies of development that offer reasonable and equitable opportunities to participate to all members of society, both Indians and non-Indians' (Maybury-Lewis 1984, p. 134).

A number of approaches to the sustainable development he advocates have of course now been made. There is increasing interest in the creation of sustainable agro-ecosystems in rainforest areas, for example (Bishop 1978, Altieri *et al.* 198), and in the problems and prospects of agroforestry (Winterbottam and Hazelwood 1987). Agroforestry appears to offer almost the ideal sustainable development technology, providing sources of fuelwood (through farm and community forestry), increasing agricultural productivity (for example through enhanced soil fertility in alley cropping), by helping small farmers (because it is essentially a micro-scale technology) and generating employment and income (Winterbotton and Hazelwood 1987). Murdia argues that in Gujarat a series of initiatives has now been taken to attempt to integrate forestry and the needs of tribal people (Murdia 1982). Budowski suggests that in Central America and the Carribbean multiple use and agroforestry seem to offer hopeful models for future forest management that will take the needs of existing local populations into account.

There are clearly many common interests between conservation of nature and the protection of the interests of tribal people, both at the level of preservationist ideology and at the pragmatic level that the wildlife of a peopled ecosystem depends on the way those people work the environment, and hence on the ways they are themselves influenced by government development policy. However, Seeger (1982) warns that nature conservation and the interests of disadvantaged groups must be kept distinct. He argues in the context of the Brazilian Amazon that the way Indians relate to their environment changes over time as populations grow and inevitable cultural changes take place. The romantic belief that the Indians are a 'natural' part of the ecosystem has led to proposals for the establishment of multiple-use reserves, creating national parks for Indians and allowing Indians in national parks. Seeger argues that this is unworkable, that 'where resources are limited, the conflict between Indians and forest management has no real solution' (Seeger 1982, p. 188), if only because of the proven capacity of others to use Indians to exploit park resources.

Perhaps, therefore, conservation and the indigenously controlled development of rural peoples are not wholly compatible. Seeger argues that although land designation is a valuable strategy for both, separate territories are needed. In this he is supported by biologists concerned about the preservation of wildlife. For example, while Johns discusses at length the possibility of achieving wildlife objectives in areas of rainforest cut by selection felling, he stresses that this 'should not be regarded as an alternative to maintaining primary forest areas' (Johns 1985, p. 370), for these alone will contain the full diversity of species. Multiple land use is therefore a pragmatic second best. It is valuable primarily where there is no political will to preserve the rainforest.

The survival of the Indians of Amazonia depends on permanent guaranteed rights to their lands, continued health and education assistance and technical advice to avoid the mistakes made by others in the rainforests. They have no need of nature conservation. Conservationists want National Parks set aside without development, although in addition they also desire other areas managed in a benign way, the 'anthropological reserves' of the IUCN Commission on National Parks and Protected Areas (IUCN Commission 1984). The primary objectives of conservationists are protected areas and species.

Development, even if packaged as 'sustainable development' is attractive chiefly as a secondary strategy where it promotes their primary objective. Where conservationists have little chance of commanding the kind of resources they desire, they are willing to push those kinds of development (which they call sustainable development) least damaging to wildlife. In doing so, they may well align themselves with the needs and aspirations of indigenous groups. Both interests have much to gain from collaboration that increases power to withstand certain kinds of development action, whether from corporate capital or the state. Nature conservationists and Indian rights activists may therefore make effective political partners, locally, nationally, and in the international field where environmental pressure groups have valuable experience and resources.

9

Green Development: Reformism or Radicalism?

Global environmental action and development policy

In the 1980s, debate about the environmental implications of development decisions has intensified and spread with the growth of international environmental action. The conservation campaigns of environmental pressure groups within industrialized countries have been vociferous and increasingly effective in influencing the actions of banks and aid agencies. This phenomenon, with the wider 'greening' of politics, is now an important focus for political scientists working within the industrialized world, particularly in the USA and Europe (Rudig and Lowe 1986, 1989).

The global reach of such environmental action has been less widely remarked upon, but it is nonetheless significant. In part this is the result of the rise of concern about international and global problems such as acid rain, the greenhouse effect, the ozone layer and Antarctica. However, the 1980s have also seen environmental organizations such as Friends of the Earth (FoE) broaden their concerns and campaigns to embrace specifically Third World issues such as deforestation and desertification. In a number of cases, this concern for the environment within Third World countries has been accompanied by the development of international networks of non-governmental environmental groups, and even federations. FoE, for example, now links thirty-five national organizations, a number of them in the Third World. The context of their protests can be rather different from that of Europe or North America. Three members of Sahabat Alam Malaysia were arrested in 1987 following a campaign of protests about rainforest logging in Borneo and the dumping of radioactive waste (Friends of the

Earth 1987). Experience of campaigning in such countries is still small, but it is growing rapidly.

Despite the growth of Third World environmental groups, the most action by environmental pressure groups continues to be within the First World. The focus for this action is reformist, particularly the modification of the policies of aid agencies. Environmental groups have run successful campaigns to persuade aid agencies and banks to modify their environmental policies and practices. In this they have been able to enlist the example of non-governmental development organizations that have themselves moved towards new and 'greener' approaches to development. The most notable achievement of this pressure has been the supposed 'greening' of the World Bank (Holden 1987, 1988, Goldsmith 1987), and it is clear that pressure of this sort has played a major part in the wider greening of development thinking in the late 1980s.

Environmental reformism: the example of rainforest environments

Sustainable development thinking contains both ecocentric and technocentric responses to the threats of development to environment and people. The former are radical and demand fundamental change in political economic structures. The latter are pragmatic, involving technical and implementable steps towards the reform of development practice. It is these technocentric and reformist approaches that have dominated the work of First World environmental pressure groups.

One of their most significant targets has been the development and management of rainforest environments. The notion of sustainable management of tropical timber resources has been an important policy goal. A series of studies have offered guidelines for development in rainforest regions (e.g. Poore 1976, Poore and Sayer 1987), and there is now a growing literature on attempts to transform forest management policies in Third World countries to include the designation of forest reserves, the creation of sustainable agriculture in rainforest regions, the managed exploitation of natural forests and the restoration of logged and degraded forest lands (see for example Gradwohl and Greenberg 1988). Repetto (1987) makes a plea for rational policies aimed at creating a renewable resource in tropical forests. Ideas of this kind have been promoted by organizations such as the World Resources Institute, a powerful and wealthy environmental think tank based in Washington, DC. In 1985 it contributed to the *Tropical Forestry Action Plan* with FAO, UNDP, the World Bank and IUCN, released at the time of the

World Forestry Congress in Mexico (Poore and Sayer 1987). This argued for increased investment in the forestry sector. It sought to make sustainable forestry in the humid tropics a reality, by doubling spending on timber and fuelwood plantations. Its target, however, was unreservedly industrial forestry, and environmentalists argued that it would neither promote the protection of remaining rainforest areas nor protect the interest of rainforest people. It was, they said, a 'top-down' approach to preserving forests (Caufield 1987).

Those eager to promote rainforest conservation have become aware that their goal requires great pragmatism. This is true both at a policy level and on the ground. One approach adopted is to explore ideas of multiple land use (Johns 1985, Osemiebo 1988). Reserves in Bendel State, Nigeria, are already penetrated by farmers and hunters, and their exclusion in the interests of wildlife conservation (even if desirable) would be politically impossible (Osemiebo 1988). As a result, wildlife conservation strategies have to be designed around existing patterns of logging and land clearance (Osemiebo 1988). A similar approach is to seek to modify regional planning in rainforest areas such as the Amazon to foster development which is both 'environmentally adaptive and economically profitable' (Eden 1978, p. 401). Another approach is to develop pragmatic arguments for rainforest conservation based on the possible economic value of genetic resources, for example in the pharmaceutical industry (Myers 1984, Cassels 1987). This approach was a central plank of the Earthlife/Bioresources initiative in the Korup National Park declared by Cameroon in the mid-1980s, and the approach is also taken by Asibey and Owusu (1982) in their justification of the need for high forest National Parks in Ghana.

Behind the pragmatism of environmental groups lies the reality of lobbying and pressure. Campaigns have not only been directed at aid agencies, but also commercial interests. The FoE Rainforest Campaign launched in 1985, for example, involved a boycott on high street outlets using tropical timber obtained through clearfelling or non-replacement selective logging. Obviously the importance of logging as a cause of forest clearance varies in different places (see Chapter 6), but the importance of First World markets for timber means that such pressure could in places have significant impacts (Oldfield 1988). A campaigning objective was the adoption of a code of conduct by UK and EC timber traders, under which they would only stock timber from concessions that have a government-approved management plan which stipulates post-logging management, where annual timber extraction does not exceed the concession's sustainable yield and where logging impacts are minimized by sympathetic extraction methods. Under the code of conduct traders should not stock wood from plantations established in virgin forest areas, and they must label country and concession of origin of wood products. Furthermore, traders would be asked to devote 1 per cent of profits towards a fund to promote the sustainable use of rainforest.

One of the foci of FoE pressure was the International Tropical Timber Organization (ITTO). This was formed in 1983 following the UN Conference on Tropical Timber held under the auspices of the UN Conference on Trade and Development (UNCTAD), and a huge amount of behind the scenes wrangling (Johnson 1985). The ITTO is a joint organization between producing and consuming countries, with votes on the Tropical Timber Council divided equally. It has been dogged by financial shortages because member countries have been slow to pay their dues. There were complaints of under-funding at its first meeting in Yokahama in March 1987. The ITTO is different from other commodity agreements because it specifically promotes reforestation and national polices of sustainable utilization and conservation. Its aims are broad, embracing the improvement of tropical forest management, the improvement of marketing and distribution of tropical timber, and the promotion of wood processing in the producing countries. Despite its problems, some observers suggest that the ITTO does offer an opportunity for conservationists, whereby they could promote conservation 'in the maximum degree of harmony with the tropical timber trade that is consonant with sound conservation objectives' (Johnson 1985, p. 44).

Beyond reformism: water resources

Parallel attempts have taken place to reconcile development and environment in the context of dam construction. The adverse environmental impacts of river control on downstream fisheries and riparian agriculture are now reasonably widely recognized. Scudder (1980, 1988) argues that these can be minimized by appropriate dam design and operation. He suggests that the simulation of the seasonal flood peak would make downstream production possible and also allow cultivation in the drawdown zone of the reservoir. In the case of a hydroelectric dam, management in this way would offset the many costs to downstream producers which need to be taken into account, and also open up a new resource for reservoir evacuees. In the case of dams built for flood control and irrigation, this form of management would reduce downstream flooding losses and allow further development to be piecemeal and locally instigated and managed, thus avoiding the high costs of centrally planned large scale irrigation. Gross benefits might be smaller, for example through reduced power generation or slower expansion of irrigation, but the cost/benefit ratio would improve considerably. There would, of course, be problems, for example in management. However, as Scudder points out, it is time that attention was paid to the management of tropical river basins rather than simply their 'development', seen as a one-off process.

The principle of using a river channel as a supply canal for irrigation is fairly familiar, and it is a relatively small step from that to seeing the floodplain as part of a water-use system which has certain minimum requirements for water supply in terms of duration and height of flows. There will be data shortages once hydrologists attempt these kind of tasks, but there is no particular difficulty to them. Indeed, Kabuzya *et al.* (1980) discuss exactly the possibility that Scudder outlines in the context of the use of water dammed on the Rufiji River in Tanzania for downstream inundation cropping.

In Scudder's formulation, river basin development is transformed from a closely-directed and externally-imposed blueprint of future development based on large scale projects to a more open-ended, flexible and diverse picture of locally-initiated smaller-scale projects. While his aim (or part of it) is to offer dam builders and river basin planners a practical alternative development model that can be implemented using existing planning frameworks, the implications of his suggestions are more fundamental. They start to challenge the whole established 'development from above' model of development planning. This 'presumes an eventually monolithic and uniform concept of development, value system and human happiness which automatically or by policy intervention will spread over the entire world' (Stohr 1981, p. 41). Stohr offers a critique of this approach, and an alternative. This is the 'development from below', the 'bottom-up and periphery-inward' development paradigm (p. 39) which plays an important role in sustainable development thinking (see Chapter 3). Development from below is an approach, not a package. The approach suggests that for success, developments must be not only innovative and research-based, but locally conceived and initiated, flexible, participatory and based on a clear understanding of local economics and politics. It is also often suggested that projects should be small in scale.

The failure of high-technology and large-scale development projects which are externally-imposed on rural communities is a common theme in environmental critiques of development. Such critiques often focus attention instead on small-scale projects (for example Timberlake 1985), developing a version of the argument that 'small is beautiful' (Schumacher 1973). This suggests that small-scale projects minimize adverse environmental impacts and maximize economic benefits, with less waste on vast management bureaucracies. It can also be argued that they make best use of the energies and talents of participants. However, few studies yield hard evidence that small-scale schemes 'work' very much better than their large-scale equivalents. There is some evidence that they exhibit similar problems. While these problems, like the schemes themselves, may be smaller in scale, they can still be costly, time-consuming and difficult to solve. Therefore, while it is attractive to be able to reduce the rather rambling ideas of sustainable development to simple rules of thumb, there are undoubted risks attached to seeing slogans such as 'small is beautiful' as something which can be implemented in a direct or simple way.

The debate about the benefits or disbenefits of small-scale projects is particularly acute in the context of African irrigation. The failings of large-scale irrigation in Africa are well known (see for example Wallace 1981, Adams and Grove 1984). Perhaps unsurprisingly, there has been increasing interest in the 1980s in small-scale irrigation (SSI) in sub-Saharan Africa (Moris and Norman 1984, Underhill 1984a, b, Adams and Carter 1987). However, the experience of small-scale irrigation sectors suggests that the scale of irrigation schemes may not be particularly important. In Kenya, for example, several small-scale irrigation schemes were established in Turkana and the North East Province in the 1960s (Hogg 1987b), with inputs from the Kenyan government, and a mixture of non-governmental organizations and international agencies. Expansion of irrigation development in Turkana began in 1979 under a UNDP/FAO project.

Although to an extent these schemes did exhibit a low-cost and low-technology approach to development, if only because of financial shortage, none of them was particularly successful. They involved mechanized cultivation, centrally controlled water distribution and cropping. Their performance was poor:, with problems of flooding, water shortage and salinity. The area irrigated in fact shrank over the project period. Plot sizes were too small, yields were low and farm incomes were well below target. The Turkana schemes grew out of measures for famine relief rather than enhanced production. Of some 8,000 Turkana in famine camps in the 1970s, 700 were settled on the various schemes, but their nutritional state was worse than that of those who had remained as pastoralists. The use of tractors for land preparation was far too costly, and the schemes required a subsidy to run between 1979 and 1983. Irrigation was obviously an inappropriate response to the destitution of pastoralists. The total development costs of the Turkana cluster schemes were about fifteen times the cost of restocking pastoralists, or the equivalent of famine relief for 200 years (Adams forthcoming).

Small-scale irrigation in Kenya appears to have the same inappropriate technology, the same high development and management costs, low yields, low settler incomes and poor economic returns as large-scale irrigation (Adams forthcoming). The actual size or scale of the irrigation schemes themselves seems rather unimportant. Enthusiasm for 'small-scale' approaches to irrigation obviously need to be tempered with considerable caution. There are no magical formulae for achieving sustainable development, and no short-cuts to policy formulation.

Pitfalls await those who expect a quick and unproblematic switch from 'unsustainable' to 'sustainable' development projects simply by altering the style of development planning, the nature of consultation with affected people and the scale of projects. The speed and scale of the endorsement of the terminology of sustainable development means that there are few examples to demonstrate this point (although see Conroy and Litvinoff 1988). Chambers (1988) probably comes as close as anybody to offering guidelines for sustainable development. His starting point is the principle of 'sustainable livelihood security'. He defines livelihood as 'adequate stocks and flows of food and cash to meet basic needs'; security he defines in terms of ownership of (or access to) resources and income-earning activities, embracing the need for reserves and assets to offset risk (p. 1). In order to achieve this, he suggests that five major lessons can be learned from experience. The first is the need to adopt a learning approach rather than a blueprint approach; the second is simply that peoples' priorities should be put first. The third is the need for a long-term view of the security of rights for the poor; without them the conservationists' dream of 'wise use' of resources must remain a mirage. His fourth lesson is that it is necessary to start with self-help, and contributions from participants. Lastly, he stresses the importance of the calibre, commitment and continuity of staff involved in development.

Complex and intractable problems can arise with developments which seem on the surface to have many of the desirable characteristics of sustainable development. Some indication of their nature is provided by the experience of small scale water resource development in the Sokoto Valley in Nigeria undertaken between 1917 and 1921 (Adams 1987). Rice production in the Sokoto valley was seriously affected by a series of years of drought in the second decade of this century, which reduced flooding. In response to the serious economic hardship this caused, the colonial District Officer had a series of small channels dug at trifling cost by the Sokoto Native Administration between the river and large closed pools in the floodplain which usually filled with water. Banks were built to retain water. Despite certain technical problems, as channels were scoured out by floodwaters while others failed to fill, major benefits in terms of rice cultivation, cattle watering and fishing were reported. In many ways this seems an ideal of sustainable development. It was low cost, involved simple technology, was stimulated by expressed local need and developed each year on the basis of experience. The technical problems which did occur could probably have been solved relatively easily had the equivalent of an Intermediate Technology engineer been available.

However, this success was illusory and brief. By 1921 this experiment in Sokoto was being dismissed as a failure. In part this was the result of the cloud under which the District Officer concerned left the Province, but there were also complaints from local village heads that the engineering works caused excessive and untimely flooding of crops and damaged fishing interests. By 1921, those banks that were not washed away were broken down. Behind the technical failures in Sokoto lay political rejection of the project, expressed through the village heads. In part this was because while some farmers gained from the banks and channels, others lost. Furthermore, although small in scale compared to the dam and irrigation developments which were undertaken on the Sokoto in the 1970s (Wallace 1980, Adams 1985b), the Sokoto project of the colonial period did not grow from within any of the communities affected. Certainly, it was triggered by local complaints, but its form was conceived by an outside 'expert' (although technically in fact he was not very expert) and it was executed by local government. The project's 'green' credentials are seen, on closer inspection, to be in large part illusory. It fulfils few of Chambers' requirements for a sustainable development project.

Green development: reformism or radicalism?

There is no magic formula for sustainable development. Despite the enthusiastic rhetoric, the technical guidelines and the lauded 'greening' of development agencies, there is no easy reformist solution to the dilemma and tragedy of poverty. Behind the slogans about environment and development lies the 'hard process' of development itself, wherein choices 'are indeed cruel' (Goulet 1971, p. 326). Development ought to be what human communities do to themselves. In practice, however, it is what is done to them by states and their bankers and 'expert' agents, in the name of modernity, national integration, economic growth or a thousand other slogans.

Fundamentally, it is this reality of development, imposed, centralizing and often unwelcome, that the 'greening' of development challenges. It throws attention back on the ethical questions that underlie the idea of development itself. It recognizes that societies are 'developing' whether or not they are the target of some specific government 'development' scheme. In practice in the Third World (as elsewhere) ideas, culture and the nature of society are in flux. Farming practice, production system, economy, are all sucked into the whirlpool of the world economy to some extent, moving in response to the pull of capital. There is no 'real development' to be reached for that escapes this, and sustainable development is no magic bridge by which it can be attained.

Part of the limitation of the sustainable development thinking and the reformist technical guidelines discussed in this book is their failure to address political economy. Without a theory of how the world economy works, and without theories about the relations between people, capital and state power, sustainable development thinking - and most conservation action - is locked within a limited compass. In practice, power and the initiative for development planning are highly centralized, globally and nationally. Development planning is centralized and imposed, as is much conventional conservation, although both seek to involve (and coopt) local interests and both believe they are operating in the interests of some notional wider constituency. Development initiatives and the context of aid-giving and project formulation have to be understood in the way the world economy functions. Pollution and environmental degradation in the Third World has to be understood in terms of its relations to the urban, industrial cores of the First World (Chambers 1988, p. 15).

Conservation is similarly dominated by global concentrations of wealth and power, and centralized decision-making. Despite the well-meaning rhetoric of environmentalists advocating sustainable development, it is not the Third World that stands to gain most from scientific advances based on the exploitation of wild genetic resources, but the industrialized economies of the North. The arrival of self-styled 'green capitalists' (Elkington and Burke 1987) does nothing to reduce the asymmetry and entrenched nature of such relations. Similarly, it is not the rural poor who will gain most from the designation of national parks, but the rich consumer in the industrialized North with leisure and wealth to be a tourist in the Third World (Cartwright 1989). The principle of directing First World resources to Third World conservation has long been important in organizations such as the World Wide Fund for Nature (formerly World Wildlife Fund), and the notion that the one who pays the piper calls the tune is being greatly extended by new ideas of 'debt for nature' swap's (Cartwright 1989). These arise from the exposure of First World Banks to Third World debt, and their willingness to sell off those debts at a discount to conservation organizations, who use them to bargain for expenditure on conservation in local currency (Ditlev-Simonson 1987).

The green alternative in development is therefore not simply about reforming environmental policy, it also issues a challenge to the very structures and assumptions of development. It is also, first and foremost about poverty and human need, what Chambers (1988) calls the principle of 'sustainable livelihood security'. He defines livelihood as 'adequate stocks and flows of food and cash to meet basic needs'; security he defines in terms of ownership of (or access to) resources and income-earning activities, embracing the need for reserves and assets to offset risk. (Chambers 1988, p. 1). Debates about the mechanisms and dynamics of development have tended to obscure its ethical basis, but Chambers argues strongly that there are moral as well as practical reasons for putting poor people first in development planning. Goulet suggests that the 'shock of underdevelopment' can only be overcome by creating 'conditions favourable to reciprocity', in which 'stronger partners ... offset the structural vulnerability of weaker interlocuters by being themselves rendered politically, economically and culturally vulnerable' (Goulet 1971, p. 328). The feasibility of such a vision can be debated, but the extent of its challenge to reformist tinkering with environmental aspects of development policy are clear.

Green development focuses on the rights of the individual to choose and control his or her own course for change, rather than having it imposed. The green agenda is therefore certainly radical, but it is also open-ended, flexible, diverse. Green development is almost a contradiction in terms, not something for which blueprints can be drawn, not something easily absorbed into structures of financial planning, or readily coopted by the state. It shares the 'politically treacherous' characteristics of O'Riordan's concept of sustainability (O'Riordan 1988, p. 30), and is subject to the very real tensions within competing ideologies of technocentric and ecocentric environmentalism (O'Riordan 1988, Turner 1988b). It is something that very often emerges in spite of, rather than as a direct result of, the actions of development bureaucracies. Green development programmes must start from the needs, understanding and aspirations of individual people, and must work to build and enhance their capacity to help themselves. As Chambers comments, 'The poor are not the problem, they are the solution' (Chambers 1988, p. 3).

This view of the desirable characteristics of development is counter-intuitive, even subversive. It is most often found among those who are recipients rather than creators of development projects. It is very much the view of the development process 'from below', looking up at what is descending from the modernizing state. In her 'letter to a young researcher', Adrian Adams describes this view with a terrible clarity. She argues that the critique of radical dependency theorists is not radical enough. They are no more able than Senegalese development bureaucrats to acknowledge what she calls the 'existence' of the village of Jamaane. But Jamaane does exist, 'here and now, having a past and a future because it is alive today' (Adams 1979, p. 477). Furthermore, there are people there with a legitimate and informed view of the development process.

Jamaane existed, and development was already happening in the Senegal Valley, created and fuelled by the interventions of successive developers. In the rural Third World as anywhere else, people live real lives. Culture, society and economy are dynamic, complex and unpredictable. 'Development' based on programmes and policies externally conceived and externally imposed is unlikely to 'work' in the economists' sense, and is likely to have adverse effects that will worry the ecologist. Such problems can certainly be tackled by promoting 'sustainable development', but this is just tinkering with the edge of the problem.

'Green' development is not about the way the environment is managed, but about who has the power to decide how it is managed. Its focus is the capacity of the poor to exist on their own terms. At its heart, therefore, 'greening' development involves not just a pursuit of ecological guidelines and new planning structures, but an attempt to redirect change to maintain or enhance the power of the poor to survive without hindrance and to direct their own lives. 'Sustainable development' is the beginning of a process, not the end.

References

Abalu, G.O.I. (1977) 'A note on crop mixtures under indigenous conditions in Northern Nigeria', *Samaru Research Bulletin*, No. 276.

Abel, N. and Stocking, M. (1981) 'The experience of underdeveloped countries', pp. 253-95 in T. O'Riordan and W.R.D. Sewell (eds) *Project Appraisal and Policy Review*, Wiley, Chichester.

Abelson, P. (1979) *Cost Benefit Analysis and Environmental Problems*, Saxon House, Farnborough.

Ackerman, W.C., White, G.F. and Worthington, E.B. (eds) (1973) *Man-Made Lakes: their problems and environmental effects*, Geophysical Monograph 17, American Geophysical Union, Washington, DC.

Adams, A. (1979) 'An open letter to a young researcher', *African Affairs* 38 (313): 451-79.

Adams, M.E. (1986) 'Merging relief and development: the case of Turkana', *Development Policy Review* 4:314-24.

Adams, W.M. (1982) 'Third World EIA: a whole new ball game', *ECOS: A Review of Conservation* 3: 30-5.

Adams, W.M. (1985a) 'The downstream impacts of dam construction: a case study from Nigeria', *Transactions of the Institute of British Geographers N.S.* 10: 292-302.

Adams, W.M. (1985b) 'River basin planning in Nigeria', *Applied Geography* 5: 292-308.

Adams, W.M. (1985c) 'River control in West Africa', pp. 177-228 in A.T. Grove (ed.) *The Niger and its Neighbours: environment, history and hydrobiology, human use and health hazards of the major West African rivers*, Balkema, Rotterdam.

Adams, W.M. (1986) 'Traditional agriculture and water use, Sokoto Valley, Nigeria', *Geographical Journal* 152: 30-44.

Adams, W.M. (1987) 'Approaches to water resource development, Sokoto Valley, Nigeria: the problem of sustainability', in D.M. Anderson and R.H. Grove (eds) *Conservation in Africa: people, policies and practice* Cambridge University Press, Cambridge.

Adams, W.M. (1988a) 'Rural protest, land policy and the planning process on the Bakolori Project, Nigeria', *Africa* 58: 315-36.

Adams, W.M. (1988b) 'Irrigation and innovation: small farmers in the Sokoto Valley, Nigeria', pp. 83-94 in J. Hirst, J. Overton, B. Allen and Y. Byron (eds) *Small Scale Agriculture*, Commonwealth Geographical Bureau and the Department of Human Geography, Research School of Pacific Studies, Australian National University, Canberra.

Adams, W.M. (forthcoming) 'How beautiful is small? Scale, control and success in Kenyan irrigation', *World Development*.

Adams, W.M. and Anderson, D.M. (1988) 'Irrigation before development: indigenous and induced change in agricultural water management in East Africa', *African Affairs* 87: 519-35.

Adams, W.M. and Carter, R.C. (1987) 'Small scale irrigation in sub-Saharan Africa', *Progress in Physical Geography* 11: 1-27.

Adams, W.M. and Grove, A.T. (eds) (1984) *Irrigation in Tropical Africa: problems and problem-solving*, Cambridge African Monographs No. 3, University of Cambridge African Studies Centre, Cambridge.

Adams, W.M. and Grove, A.T. (1987) 'The implications of climatic variability for river regulation', pp. 5-17 in J. Seeley and W.M. Adams (eds) *Environmental Issues in African Development Planning*, Cambridge African Monographs No. 9, University of Cambridge African Studies Centre, Cambridge.

Adams, W.M. and Hollis, G.E. (1989) *Hydrology and Sustainable Resource Development of a Sahelian Floodplain Wetland*, Hadejia-Nguru Wetlands Project, University College, London.

Adams, W.M. and Rose, C.I. (eds) (1978) *The Selection of Reserves for Nature Conservation*, Discussion Papers in Conservation No. 20, University College London.

Adamson, R.S. (1938) *The Vegetation of South Africa*, British Empire Vegetation Committee, London.

Addo, H., Amin, S., Aseniero, G., Frank, A.G., Friberg, M., Frobel, F., Heinrichs, J., Hettne, B., Kreye, O. and Seki, H. (1985) *Development as Social Transformation: reflections on the global problematique*, Hodder and Stoughton, Sevenoaks, for the United Nations University.

Adeniyi, E.O. (1970) 'The Kainji Dam: an exercise in regional planning', *Regional Studies* 10: 233-43.

Adeniyi, E.O. (1973) 'The impact of change in river regime on economic activities below Kainji Dam', pp. 169-77 in A.L. Mabogunje (ed.) *Kainji Lake Studies Volume II*, Nigerian Institute of Social and Economic Research, Ibadan.

Africa Confidential (1986) 'Kenya: a sudden chill', *Africa Confidential* 27(8): 1-3.

Africa Confidential (1987) 'Kenya: electric shock', *Africa Confidential* 28(7): 7.

Ahmed, Y.J. and Sammy, G.K. (1985) *Guidelines to Environmental Impact Assessment in Developing Countries*, Hodder and Stoughton, Sevenoaks.

Alabaster, J.S. (1970) 'River flow and upstream movement and catch of migratory salmonids', *Journal of Fisheries Biology* 2: 1-13.

Allan, W. (1965) *The African Husbandman*, Oliver and Boyd, Edinburgh.

Allen, D.E. (1976) *The Naturalist in Britain*, Penguin, Harmondsworth.

Allen, D.E. (1986) *The Botanists: a history of the Botanical Society of the Brtitish Isles through a hundred and fifty years*, St Pauls Bibliographies, Winchester.

Allen, F. (1983) 'Natural resources as natural fantasies', *Geoforum* 14: 243-7.

Allen, J.C. and Barnes, D.F. (1985) 'The causes of deforestation in developing countries', *Annals of the Association of American Geographers* 75: 163-84.

Allen, R.N. (1973) 'The Anchicayá Hydroelectric Project in Colombia: design and sedimentation problems', pp. 318-42 in M.T. Farvar and J.P. Milton (eds) *The Careless Technology: ecology and international development.* Stacey, London.

Allen, R. (1980) *How to Save the World: strategy for world conservation.*, Kogan Page, London.

Alsopp, R., Hall, D. and Jones, T. (1985) 'Fatal attraction for the tsetse fly', *New Scientist* 7 November: 39-43.

Altieri, M.A., Letourneau, D.K. and Davis, J. R. (1984) 'The requirements of sustainable agro-ecosystems', pp. 175-89 in G.K. Douglass (ed.) *Agricultural Sustainability in a Changing World Order*, Westview, Boulder, Colo.

Ambio (1979) 'Review of Environmental Development', *Ambio* 8: 114-15.

Amin, S. (1985) 'Apropos the "green" movements', in H. Addo *et al.* *Development as Social Transformation: reflections on the global problematique*, Hodder and Stoughton, Sevenoaks, for the United Nations University.

Anadu, P.A. (1987) 'Progress in the conservation of Nigeria's wildlife', *Biological Conservation* 41: 137-251.

Anderson, C.W. (1912) *Forests of British Guiana* (2 vols), Department of Lands and Mines, Georgetown, British Guiana.

Anderson, D.M. (1984) 'Depression, dust bowl, demography and drought: the colonial state and soil conservation in East Africa during the 1930s', *African Affairs* 83: 321-44.

Anderson, D.M. and Grove, R.H. (eds) (1987) *Conservation in Africa: people, policies and practice.* Cambridge University Press, Cambridge.

Anderson, D.M. and Millington, A. (1987) 'Political ecology of soil conservation in anglophone Africa', in A.C. Millington, S. Mutiso and J.A. Binns (eds) *African Resources: Volume II*, Department of Geography, University of Reading.

Andrae, G. and Beckman, B. (1985) *The Wheat Trap: bread and underdevelopment in Nigeria*, Zed Press, London.

Andrews, D.J. (1975) Intercropping with sorghum, *Samaru Research Bulletin* No. 238.

Apeldoorn, G.J. van (1980) *Perspectives on Drought and Famine in Nigeria*, Allen and Unwin, Hemel Hempstead.

Arvelo-Jimeniz, N. (1984) 'The politics of cultural survival in Venezuela: beyond indigismo', pp. 105-26 in M. Schmink and C.H. Wood (eds) *Frontier Expansion in Amazonia*, University of Florida Press, Gainesville, Fla.

Aseniero, G. (1985) 'A reflection on developmentalism: from development to transformation', pp. 48-85 in H. Addo *et al.* *Development as Social Transformation: reflections on the global problematique*, Hodder and Stoughton, Sevenoaks, for the United Nations University.

Ashby, E. (1980) 'Retrospective Environmental Impact Assessment', *Nature* 288: 28-9.

Asibey, E.O.A. and Owusu, J.G.K. (1982) 'The case for high-forest national parks in Ghana', *Environmental Conservation* 9: 293-304.

Atkinson, T.R. (1973) 'Resettlement programmes in the Kainji Lake Region', in A.L. Mabogunje (ed.) *Kainji: a Nigerian man-made lake; Kainji Lake Studies Volume 2: Socio-economic conditions*, Nigerian Institute for Social and Economic Research, Ibadan.

Attfield, R. (1983) 'Christian attitudes to nature', *Journal of the History of Ideas* 44: 369-86.
Attwell, R.I.G. (1970) 'Some effects of Lake Kariba on the ecology of a floodplain of the mid-Zambezi Valley of Rhodesia', *Biological Conservation* 2: 189-96.
Aubréville, A. (1949) *Climats, Forêts et Désertification de l'Afrique Tropicale*, Société d'Edition Géographiques, Maritimes et Coloniales, Paris.
Bahro, R. (1978) *The Alternative in Eastern Europe*, New Left Books, London (republished 1981, Verso, London).
Bahro, R. (1982) *Socialism and Survival*, Heretic, London.
Bahro, R. (1984) *From Red to Green: interviews with New Left Review*, Verso, London.
Baldwin, J.H. and Sachs, A.B. (1984) 'The origin, evolution and status of the International Society for Environmental Education (ISEE)', *Environmental Conservation* 11: 313-17.
Baldwin, K.D.S. (1957) *The Niger Agricultural Project: an experiment in African development*, Blackwell, Oxford.
Balon, E.K. and Coche, A.G. (eds) (1974) *Lake Kariba: a man-made tropical ecosystem in Central Africa*, Monographiae Biologicae 24, Junk, The Hague.
Baluyut, E.A. (1985) 'The Agno River Basin (The Philippines)', pp. 15-54 in T. Petr (ed.) *Inland Fisheries in Multi-purpose River Basin Planning and Development in Tropical Asian Countries: three case studies*, FAO Fisheries Technical Paper 265, Food and Agriculture Organisation, Rome.
Bardach, J.F. (1973) 'Some ecological consequences of Mekong River development plans', in M.T. Farvar and J.P. Milton (eds) *The Careless Technology: ecology and international development*, Stacey, London.
Barker, D. and Spence, B. (1988) 'Afro-Carribbean agriculture: a Jamaican Maroon Community in transition', *Geographical Journal* 154: 198-208.
Barker, J. (ed.) (1984) *The Politics of Agriculture in Tropical Africa*, Sage, Beverly Hills.
Barnett, T. (1980) 'Development: black box or Pandora's box?', pp. 306-24 in J. Heyer, P. Roberts and G. Williams (eds) *Rural Development in Tropical Africa*, Macmillan, London.
Barney, G.O. (1982) *The Global 2000 Report to the President*, Council on Environmental Quality and Department of State, Penguin Books, New York.
Barrow, C.J. (1981) 'Health and resettlement consequences and opportunities created as a result of river impoundment in developing countries', *Water Supply and Management* 5: 135-50.
Barrow, C.J. (1987) 'The environmental effects of the Tucurí Dam on the middle and lower Tocatins River Basin, Brazil', *Regulated Rivers* 1: 44-60.
Bass, S.M.J. (1988) 'National Conservation Strategy, Zambia', pp. 186-91 in C. Conroy and M. Litvinoff (eds) *The Greening of Aid: sustainable livelihoods in practice*, Earthscan, London.
Bates, R.H. (1983) *Essays in the Political Economy of Rural Africa*, Cambridge University Press, Cambridge.
Batisse, M. (1975) 'Man and the Biosphere', *Nature* 256: 156-8.
Batisse, M. (1982) 'The biosphere reserve: a tool for environmental conservation and management', *Environmental Conservation* 9: 101-11.

Bayliss-Smith, T.P., Bedford, R., Brookfield, H. and Latham, M. (1988) *Islands, Islanders and the World: the colonial and post-colonial experience of eastern Fiji*, Cambridge University Press, Cambridge.

Baxter, G. (1961) 'River utilisation and the preservation of migratory fish', *Proceedings of the Institution of Civil Engineers* 18: 225-44.

Baxter, R.M. (1977) 'Environmental effects of dams and impoundments', *Annual Review of Ecology and Systematics* 8: 255-84.

Beckerman, W. (1974) *In Defence of Economic Growth*, Jonathan Cape, London.

Beckman, B. (1986) 'Bakolori: peasants versus the state in Nigeria', pp. 140-55 in G. Goldsmith and N. Hildyard (eds) *The Social and Environmental Effects of Large Dams; Volume 2: Case Studies*, Wadebrige Ecological Centre, Wadebridge, England.

Beinart, W. (1984) 'Soil erosion, conservation and ideas about development: a Southern African exploration 1900-1960', *Journal of Southern African Studies* 11: 52-84.

Beinart, W. and Bundy, C. (1980) 'State intervention and rural resistance: the Transkei 1900-1965', pp. 272-315 in M. Klein (ed.) *Peasants in Africa: historical and contemporary perspectives*, Sage Publishing, Beverly Hills and London.

Bell, J. (1986) 'Caustic waste menaces Jamaica', *New Scientistt* 3 April: 33-7.

Belshaw, D. (1974) 'Taking indigenous technology seriously: the case of intercropping techniques in East Africa', *Institute of Development Studies Bulletin* 10: 24-7

Bennet, (ed.) (1978) *The Dammed: the plight of the Akawaio Indians of Guyana*, Document VI, Survival International, London.

Bernstein, H. (1979) 'African peasantries: a theoretical framework', *Journal of Peasant Studies* 6: 420-44.

Berry, L. (1984) 'Desertification in the Sudan-Sahelian region (1977-1984)', *UNEP Desertification Control Bulletin* 10: 23-8.

Bews, J.W. (1916) 'An account of the chief types of vegetation in South Africa, and notes on the plant succession', *Journal of Ecology* 4: 129-59.

Bideleux, R. (1987) *Communism and Development*, Methuen, London.

Bishop, J.P. (1978) 'Development of a sustained yield agrosystem in the upper Amazon', *Agro-ecosystems* 4: 459-61.

Biswas, A.K. (ed.) (1980) 'The Nile and its environment', *Water Supply and Management* 4: 1-113.

Biswas, A.K. and Biswas, M.R. (1976) 'Hydropower and the environment', *Water Power and Dam Construction* 28: 40-3.

Biswas, M.R. and Biswas, A.K. (1984) 'Complementarity between environment and development processes', *Environmental Conservation* 11: 35-44.

Blaikie, P. (1985) *The Political Economy of Soil Erosion*, Longman, London.

Blaikie, P. (1988) 'The explanation of land degradation in Nepal', pp. 132-58 in J. Ives and D.C. Pitt (eds) *Deforestation: social dynamics in watersheds and montane ecosystems*, Routledge, London.

Blaikie, P. and Brookfield, H. (1987) *Land Degradation and Society*, Methuen, London.

Bliss-Guest, P.A. and Keckes, S. (1982) 'The regional seas programme of UNEP', *Environmental Conservation* 9: 43-9.

Boardman, R. (1981) *International Organisations and the Conservation of Nature*, Indiana University Press, Bloomington, Ind.

Boffey P.M. (1976) 'International Biological Programme: was it worth the cost and effort?', *Science* 193: 866-8.

Bookchin, M. (1979) 'Ecology and revolutionary thought', *Antipode* 10(3)/11(1): 21-32.

Botswana Society (1976) *Proceedings of the Symposium on the Okavango Delta and its Future Utilisation*, National Museum, Gaborone, Botswana, 30 August to 2 September.

Boulding K.E. (1966) 'The economics of the coming spaceship earth', in *Environmental Quality in a Growing Economy*, (reprinted pp. 121-32 in H.E. Daly (ed.) (1973) *Towards a Steady-state Economy*. W.H. Freeman, New York)

Bovill, E.W. (1921) 'The encroachment of the Sahara on the Sudan', *Journal of the African Society* 20: 174-85.

Boyd, J.M. (1965) 'Research and management in East African Wildlife', *Nature* 208: 828-36.

Boyd-Orr, Sir J. (1953) *The White Man's Dilemma: food and the future*, George Allen and Unwin, London.

Brandt, W. (1980) *North-South: a programme for survival*, Pan, London.

Brandt, W. (1983) *Common Crisis North-South: cooperation for world recovery*, Pan, London.

Breitbart, M. (1981) 'Peter Kropotkin: the anarchist geographer', in D.R. Stoddart (ed.) *Geography, Ideology and Social Concern*, Basil Blackwell, Oxford.

Bremen, H. and de Wit, C.T. (1983) 'Rangeland productivity and exploitation in the Sahel', *Science* 221: 1341-7.

Briggs, D. (1986) 'Environmental problems and policies in the European Community', pp. 105-44 in C. C. Park (ed.) *Environmental Policies: an international review*, Croom Helm, London .

Brodie, J. and Morrison, J. (1984) 'The management and disposal of hazardous wastes in the Pacific Islands', *Ambio* 13: 331-3.

Brokensha, D.W. and Scudder, T. (1968) 'Resettlement', pp. 22-62 in N. Rubin and W.M. Warren (eds) *Dams in Africa*, Frank Cass, London.

Brokensha, D.W., Warren, D.M. and Warner, O. (eds) (1980) *Indigenous Systems of Knowledge and Development*, University Press of America, Lanham, Washington.

Brookfield, H. (1975) *Interdependent Development*, Methuen, London.

Brooks, F.T. (ed.) (1925) *Imperial Botanical Conference, London 1924: Report of proceedings*, Cambridge University Press, Cambridge.

Brown, J.G. *et al.* (eds) (1964) *World Register of Dams*, International Commission on Large Dams, Paris.

Brown, L. and Wolf, E.C. (1984) *Soil Erosion: quiet crisis in the world economy*, Worldwatch Paper 60, Washinton, DC.

Brundtland, H. (1987) *Our Common Future*, Oxford University Press, Oxford, for the World Commission on Environment and Development).

Budiardjo, C. (1986) 'The politics of Transmigration', *The Ecologist* 16: 57-116.

Budowski, G. (1982) 'The socio-economic effects of forest management on the lives of people living in the area: the case of Central America and some Carribbean countries', pp. 87-166 in E.G. Hallsworth (ed.) *Socio-economic Effects and Constraints in Tropical Forest Management*, Wiley, Chichester.

Bull, D. (1982) *A Growing Problem: pesticides and the Third World*, Oxfam Books, Oxford.

Buma, P.G, and Day, J.C. (1977) 'Channel morphology below reservoir storage projects', *Environmental Conservation* 4: 279-84.

Bunker, S.G. (1980) 'The impact of deforestation on peasant communities in the Medio Amazonas of Brazil', pp. 45-60 in V.H. Sutlive, N. Altshuler and M.D. Zamora (eds) *Where Have All the Flowers Gone? Deforestation in the Third World*, Studies in Third World Societies No. 13, College of William and Mary, Williamsburg, Va.

Bunting, A.H., Dennett, M.D., Elston, J. and Milford, J.R. (1976) 'Rainfall trends in the West African Sahel', *Quarterly Journal of the Royal Meteorological Society* 102: 59-64.

Burgess, R. (1978) 'The concept of nature in geography and Marxism', *Antipode* 10: 1-11.

Butcher, A.D. (1967) 'A changing aquatic fauna in a changing environment', *Proc. and Papers IUCN 10th Technical Meeting*. IUCN Publications new series 9: 197-218.

Butcher, D.A. (1971) *An operational manual for resettlement: a systematic approach to the resettlement problem created by man-made lakes, with special relevance for West Africa*, Food and Agriculture Organisation, Rome

Buxton, P.A. (1935) 'Seasonal changes in vegetation in the North of Nigeria,' *Journal of Ecology* 23: 134-9.

Byres, T. (1979) 'Of neo-populist pipe-dreams', *Journal of Peasant Studies* 4: 210-44.

Caldwell, L.K. (1984) 'Political aspects of ecologically sustainable development', *Enviromental Conservation* 11: 299-308.

Caldwell, M. (1977) *The Wealth of Some Nations*, Zed Press, London.

Capra, F. and Spretnak, C. (1984) *Green Politics*, E.P. Dutton, New York (republished 1985 as Spretnak and Capra, *Green Politics: the global promise*, by Paladin, London, 1985).

Carim, E., Barnard, G., Foley, G., de Silva, D., Tinker, J. and Walgate R., (eds.) (1987) *Towards Sustainable Development*, Panos Institute, London.

Carr-Saunders, A.M. (1922) *The Population Problem: a study in human evolution*, Clarendon Press, Oxford.

Carr-Saunders, A.M. (1936) *World population: past growth and present trends*, Clarendon Press, Oxford.

Cartwright, J. (1989) 'Conserving nature, decreasing debt', *Third World Quarterly* 11: 114-27

Cassels, R. (1987) 'How credible is the jungle pharmacy?' *Earthlife News* 6: 32-3.

Castleman, B. (1981) 'Double standards: asbestos in India', *New Scientist* 26 February: 522-3 .

Castri, F. di (1986) 'Interdisciplinary research for the ecological development of mountain and island areas', pp. 301-16 in N. Polunin (ed.) *Ecosystem Theory and Application*, Wiley, Chichester.

Caufield, C. (1982) *Tropical Moist Forests: the resource, the people, the threat*, Earthscan/ IIED, London.

Caufield, C. (1984) 'Pesticides: exporting death', *New Scientist* 16 August: 15-17.

Caufield, C. (1985) *In the Rainforest*, Pan Books, London.

Caufield, C. (1987) 'Conservationists scorn plans to save tropical forests', *New Scientist* 25 June: 33.

Cernea, M. (1988) *Involuntary Resettlement in Development Projects: policy guidelines in World Bank-financed projects*, Technical Paper 80, World Bank, Washington, DC.

Chambers, R. (1969) *Settlement Schemes in Tropical Africa: a study of organisation and development*, Routledge and Kegan Paul, London.

Chambers, R. (ed.) (1970) *The Volta Resettlement Experience*, Pall Mall, London.

Chambers, R. (1977) 'Challenges for rural research and development', pp. 398-412 in B.H. Farmer (ed.) *Green Revolution? Technology and change in rice-growing areas of Tamil Nadu and Sri Lanka*, Macmillan, London.

Chambers, R. (1980) 'Cognitive problems of experts in rural Africa', pp. 96-102 in J.C. Stone (ed.) *Experts in Africa*, University of Aberdeen African Studies Group, Aberdeen.

Chambers, R. (1983) *Rural Development: putting the last first*, Longman, London.

Chambers, R. (1988) 'Sustainable rural livelihoods: a key strategy for people, environment and development', pp. 1-17 in C. Conroy and M. Litvinoff (eds) *The Greening of Aid: sustainable livelihoods in practice*, Earthscan, London.

Chapman, K. (1981) 'Issues in Environmental Impact Assessment', *Progress in Human Geography* 5: 190-210.

Charney, J. (1975) 'Dynamics of deserts and drought in the Sahara', *Quarterly Journal of the Royal Meteorological Society* 101: 193-202.

Charney, J., Stone, P.H. and Quirk, W.J. (1975) 'Drought in the Sahara: a biogeophysical feedback mechanism', *Science* 187: 434-5.

Chien, N. (1985) 'Changes in river regime after the construction of an upstream reservoir', *Earth Surface Processes and Landforms* 10: 143-59.

Chilcote, R.H. (1984) *Theories of Development and Underdevelopment*, Westview Press, Boulder, Colo.

Chisholm, A. (1972) *Philosophers of the Earth: conversations with ecologists*, Sidgwick and Jackson, London.

Chisholm, N.G. and Grove, J.M. (1985) 'The Lower Volta', pp. 229-50 in A.T. Grove (ed.) *The Niger and its Neighbours: environment, history and hydrobiology, human use and health hazards of the major West African rivers*, Balkema, Rotterdam.

Christiansson, C. and Ashuvud, J. (1985) 'Heavy industry in a rural tropical ecosystem', *Ambio* 14: 122-33.

Clark, B. (1976) 'Evaluating environmental impacts', pp. 91-104 in T. O'Riordan and R. Hey (eds) *Environmental Impact Assessment*, Saxon House, Farnborough.

Clark, B.D., Chapman, K., Bisset, R., Wathern, P. and Barrett, M. (1981) *A Manual for the Assessment of Major Development Proposals*, HMSO, London.

Clarke, R. and Timberlake, L. (1982) *Stockholm Plus Ten: Promises Promises? The decade since the 1972 UN Environment Conference*, Earthscan, London.

Clarke, W.C. and Munn, R.E. (eds) (1987) *Sustainable Development of the Biosphere*, Cambridge University Press, Cambridge.

Clay, E.J. and Schaffer, B.B. (eds) (1984) *Room for Manoeuvre: an exploration of public policy in agriculture and rural development*, Heinemann, London.

Clements, F.E. (1935) 'Experimental ecology in the public service', *Ecology* 16: 342-63.

Coe, M.J., Cummings, D.H. and Phillipson, J. (1976) 'Biomass and production of large African herbivores in relation to rainfall and primary production', *Oecologia* 22: 341-54.

Colchester, M. (1985) 'An end to laughter? Hydropower projects in central India', in *An End to Laughter?*, Review 44, Survival International, London.

Cole, S. (1978) 'The global futures debate 1965-1976', in C. Freeman and M. Jahoda (eds) *World Futures: the great debate*, Martin Robinson, London.

Colinvaux, P. (1979) 'The ice-age Amazon', *Nature* 278: 399-400.

Collett, D. (1987) 'Pastoralists and wildlife: image and reality in Kenyan Masailand', pp. 129-48 in D.M. Anderson and R.H. Grove (eds) *Conservation in Africa: people, policies and practice*, Cambridge University Press, Cambridge.

Colson, E. (1971) *The Social Consequences of Resettlement*, (Kariba Studies iv), University of Manchester Press, Manchester.

Commoner, B. (1972) *The Closing Circle: confronting the environmental crisis*, Jonathan Cape, London.

Conklin, H.L. (1954) 'An ethnological approach to shifting agriculture', *Transactions of the New York Academy of Science* 77: 133-42.

Conroy, C. (1988) 'Introduction', pp. xi-xiv in C. Conroy and M. Litvinoff (eds) *The Greening of Aid: sustainable livelihoods in practice*, Earthscan, London.

Conroy, C. and Litvinoff, M. (eds) (1988) *The Greening of Aid: sustainable livelihoods in practice*, Earthscan, London.

Conwentz, H. (1914) 'On national and international protection of nature', *Journal of Ecology* 2: 109-22.

Copans, J. (1983) 'The Sahelian drought: social sciences and the political economy of underdevelopment', pp. 83-97 in K. Hewitt (ed.) *Interpretations of Calamity: from the viewpoint of human ecology*, Allen and Unwin, Hemel Hempstead.

Corbridge, S.E. (1982) 'Interdependent development? Problems of aggregation and implementation in the Brandt Report', *Applied Geography* 2: 253-65.

Corbridge, S.E. (1986) *Capitalist World Development: a critique of radical development geography*, Macmillan, London.

Cotgrove, S. (1976) 'Environmentalism and Utopia', *Sociology Review* 24: 23-42.

Cotgrove, S. (1982) *Catastrophe or Cornucopia: the environment, politics and the future*, Wiley, Chichester.

Cotgrove, S. and Duff, A. (1980) 'Environmentalism, middle class radicalism and politics', *Sociology Review* 28: 235-351.

Coughenour, M.B., Ellis, J.E., Swift, D.M., Coppock, D.L., Galvin, K., McCabe, J.T. and Hart, T.C. (1986) 'Livestock feeding ecology and resource utilisation in a nomadic pastoral ecosystem', *Journal of Applied Ecology* 23: 573-83.

Courel, M.F., Kandel, R.S. and Rasool, S.I. (1984) 'Surface albedo and the Sahel drought', *Nature* 307: 528-31.

Cox, P. (1985) *Pesticide Use in Tanzania*, ODI Economic Research Bureau, Dar es Salaam.

Crehan, K. (1985) 'Bishimi and social studies: the production of knowledge in a Zambian village', *African Affairs* 84: 89-110.

Crisp, D.T., Mann, R.H.K. and Cubby, P.R. (1983) 'Effects of regulation of the River Tees upon fish populations below Cow Green Reservoir', *Journal of Applied Ecology* 20: 371-86.

Crosby, A.W. (1986) *Ecological Imperialism: the ecological expansion of Europe* 1600-1900, Cambridge University Press, Cambridge.

Crummey, D. (ed.) (1986) *Banditry, Rebellion and Social Protest in Africa*, James Currey/Heinemann, London.

Culwick, A.T. (1943) 'New beginning', *Tanganyika Notes and Records* 15: 1-6.

Curry-Lindahl, K. (1986) 'The conflict between development and nature conservation with special reference to desertification', pp. 106-130 in N. Polunin (ed.) *Ecosystem Theory and Application*, Wiley, Chichester.

Dahlberg, K.A., Soroos, M.S., Ferau, A.T., Harf, J.E. and Trout, B.T. (1985) *Environment and the Global Arena: actors, values, policies and futures*, Duke University Press, Durham, NC.

Daly, H.E. (ed.) (1973) *Towards a Steady-state Economy*, W.H. Freeman, New York.

Daly, H.E. (1977) *Steady-state Economics: the economics of biophysical equilibrium and moral growth*, W.H. Freeman, New York.

Dasmann, R.F. (1980) 'Ecodevelopment: an ecological perspective', pp. 1331-5 in J.I. Furtado (ed.) *Tropical Ecology and Development*, International Society of Tropical Ecology, Kuala Lumpur.

Dasmann, R.F., Milton, J.P. and Freeman, P.H. (1973) *Ecological Principles for Economic Development*, Wiley, Chichester.

Davidson, F.A., Vaughan, E., Hutchinson, S.J. and Pritchard, A.L. (1973) 'Factors affecting the upstream migration of pink salmon (*Oncorhynchus gorbuscula*)', *Ecology* 24: 149-68.

Davies, B.R. (1979) 'Stream regulation in Africa: a review', pp. 113-42 in J.V. Ward and J.A. Stanford (eds) *The Ecology of Regulated Streams*, Plenum Press, New York.

Davies, B.R., Hall, A. and Jackson, P.B.N. (1972) 'Some ecological aspects of the Cabora Bassa Dam', *Biological Conservation* 8: 189-201.

Davies, B.R. and Walker, K.F. (eds) (1986) *The Ecology of River Systems*, Monographiae Biologicae 60, Junk, The Hague.

Davis, T.A.W. and Richards, P.W. (1933) 'The vegetation of Moraballi Creek, British Guiana: an ecological study of a limited area of tropical rain forest', Part I, *Journal of Ecology* 21: 350-84 .

Davos, C.A. (1977) 'Towards an Integrated Environmental Impact Assessment within a social context', *Journal of Environmental Management* 5: 297-306.

Davy, J. B. (1925) 'Correlation of taxonomic work in the Dominions and Colonies with work at home', pp. 214-34 in F.T. Brooks (ed.) *Imperial Botanical Conference, London 1924, Report of Proceedings*, Cambridge University Press, Cambridge.

Dee, N., Baker, J.K., Droby, N.L., Duke, K.M.and Fahringer, D.L. (1972) *Environmental Evaluation System for Water Resource Planning*, Battelle Columbus Laboratories, Columbus, Ohio.

Denevan, W.M. (1973) 'Development and the imminent demise of the Amazonian forest', *Professional Geographer* 25: 130-5.

Denevan, W.M. (1980) 'Swiddens and cattle versus forest: the imminent demise of the Amazonian rain forest re-examined', pp. 225-44 in V.H. Sutlive, N. Altshuler and M.D. Zamora (eds) *Where Have All The Flowers Gone? Deforestation in the Third World*, Studies in Third World Societies No. 13, College of William and Mary, Williamsburg, Va.

Denevan, W.M., Treacy, J.M., Alcorn, J.B., Padock, C., Denslow, J. and Paitan, S.F. (1984) 'Indigenous agroforestry in the Peruvian Amazon: Bora Indian management of swidden fallows', *Interciencia* 9: 346-57.

Dennett, M.D., Elston, J., and Rodgers, J.A. (1985) 'A reappraisal of rainfall trends in the Sahel', *Journal of Climatology* 5: 353-61.

Department of the Environment (1988) *Our Common Future: a perspective by the UK on the Report of the World Commission on Environment and Development*, Department of the Environment, London.

Derman, W. (1984) 'USAID in the Sahel: development and poverty', in J. Barker (ed.) *The Politics of Agriculture in Tropical Africa*, Sage, Beverly Hills.

Derman, W. and Whiteford, S. (eds) (1985) *Social Impact Analysis and Development Planning in the Third World*, Westview Press, Boulder, Colo.

Deshmukh, I. (1986) *Ecology and Tropical Biology*, Blackwell, Oxford.

Ditlev-Simonsen, C. (1987) 'Bolivia to swap debt for ecology', *Guardian* 15 July: 7.

D'Itri, P.A. and D'Itri, F.M.D. (1977) *Mercury Contamination: a human tragedy*, Wiley, Chichester.

Doughty, R. (1981) 'Environmental theology: trends and prospects in Christian thought', *Progress in Human Geography* 5: 234-48.

Douglass, G.K. (ed.) (1984) *Agricultural Sustainability in a Changing World Order*, Westview Press, Boulder, Colo.

Dregne, H.E. (1984) 'Combatting desertification: evaluation of progress', *Environmental Conservation* 11: 115-21.

Drew, E.B. (1976) 'Dam outrage: the story of the Army Engineers', *Atlantic Monthly* April: 51-62.

Dunning, M.H. (1979) 'Government-managed land settlement and resettlement in Thailand', pp. 15-26 in L. Gosling (ed) *Population Resettlement in the Mekong River Basin.* Studies in Geography No. 10, University of North Carolina Press, Chapel Hill, NC.

Eckholm, E.P. (1976) *Losing Ground: environmental stress and world food prospects,* Pergamon Press, Oxford.

Eckholm, E.P. (1982) *Down to Earth: environment and human needs,* IIED and Earthscan, London.

Eden, M. J. (1978) 'Ecology and land development: the case of Amazonian rainforest', *Transactions of the Institute of British Geographers N.S.* 3: 444-63.

Edwards, C. (1985) *Fragmented World: perspectives on trade, money and crisis,* Methuen, London.

Ehrlich, P.R. (1972) *The Population Bomb,* Ballantine, London.

Ehrlich, P.R. and Ehrlich, A.H. (1970) *Population, Resources and Environment: issues in human ecology,* W.H. Freeman, New York.

Elkington, J. and Burke, T. (1987) *The Green Capitalists: in search of environmental excellence,* Victor Gollancz, London.

Elliott, C. (1982) *Real Aid: a strategy for Britain,* Independent Group on British Aid, Oxford.

Elliott, C. (1984) *Aid is Not Enough: Britain's policies to the world's poor,* Independent Group on British Aid, Oxford.

Environmental Conservation (1982) 'Declaration: the World Campaign for the Biosphere', *Environmental Conservation* 9: 91-2.

Enzensberger, H.M. (1974) 'A critique of political ecology', *New Left Review* 8: 3-32 .

Evans, G.C. (1939) 'Ecological studies on the rainforest of southern Nigeria: II the atmospheric environmental conditions', *Journal of Ecology* 26: 436-82.

Farid, M.A. (1975) 'The Aswan High Dam development project', pp. 89-102 in N.F. Stanley and M.P. Alpers (eds) *Man-made Lakes and Human Health,* Academic Press, London.

Faulkner, O.T. and Mackie, J.R. (1933) *West African Agriculture,* Cambridge University Press, Cambridge.

Faure, H. and Gac, Y.-V. (1981) 'Will the Sahelian drought end in 1985?', *Nature* 291: 475-8.

Fearnside, P.M. (1980) 'The effects of cattle pasture on soil fertility in the Brazilian Amazon: consequences for beef production sustainability', *Tropical Ecology* 21: 125-37.

Fearnside, P.M. (1986) 'Spatial concentration of deforestation in the Brazilian Amazon', *Ambio* 15: 74-81.

Ferau, A.T. (1985) 'Environmental actors', pp. 43-67 in K.A. Dahlberg *et al.* (1985) *Environment and the Global Arena: actors, values, policies and futures,* Duke University Press, Durham, NC.

Fitter, R.S.R. (1978) *The Penitent Butchers,* Collins, London.

Fitzgerald, M. (1986) 'Kenyan dam contract provokes criticism', *Financial Times* 19 March: 7.

Flenley, J. (1979) 'The late Quaternary vegetational history of the equatorial mountains', *Progress in Physical Geography* 3: 488-509.

Flohn, H. and Nicholson, S.E. (1980) 'Climatic fluctuations in the arid belt of the "Old World" since the last glacial maximum: possible causes and future implications,' *Palaeoecology of Africa* 12: 3-21.

Folland, C.K., Palmer, T.N. and Parker, D.E. (1986) 'Sahel rainfall and worldwide sea temperatures 1901-1985', *Nature* 320: 602-7.

Forbes, D.K. (1984) *The Geography of Underdevelopment: a critical survey*, Croom Helm, London.

Ford, J. (1971) *The Role of Trypanosomiasis in African Ecology: A study of the tsetse fly problem*, Clarendon Press, Oxford.

Forrest, T. (1981) 'Agricultural policies in Nigeria 1900-1978', pp. 222-58 in J. Heyer, P. Roberts and G. Williams (eds) *Rural Development in Tropical Africa*, MacMillan, London.

Forrester, J.W. (1971) *World Dynamics*, Wright-Allen Press, Cambridge, Mass.

Fosberg, F.R. (1963) *Man's Place in the Island Ecosystem*, Bishop Museum Press, Hawaii.

Frank, A.G. (1975) *On Capitalist Underdevelopment*, Oxford University Press, Bombay.

Frank, A.G. (1980) 'North-South and East-West: Keynsian paradoxes in the Brandt Report', *Third World Quarterly* 2: 669-80.

Frank, A.G. (1981) *Crisis in the Third World*, Heinemann, London.

Frank, L. (1987) 'The development game', *Granta* 22: 231-43

Franke, R.W. and Chasin, B.H. (1979) 'Peanuts, peasants and pastoralists: the social and economic background to ecological deterioration in Niger', *Journal of Peasant Studies* 8: 1-30.

Franke, R.W. and Chasin, B.H. (1980) *Seeds of Famine: ecological destruction and the development dilemma in the West African Sahel*, Allenheld and Osman, Montclair, NJ.

Frankel, S. (1953) *The Economic Impact on Under-developed Societies: essays on international development and social change*, Blackwell, Oxford.

Franklin, J.F. (1977) 'The Biosphere Reserve Programme in the United States', *Science* 195: 262-7.

Fraser Darling, F. (1955) *West Highland Survey: an essay in human ecology*, Oxford University Press, Oxford.

Freeman, P.H. (1977) *Large Dams and the Environment: recommendations for development planning*, Report for United Nations 1977 Water Conference, IIED, London.

Friberg, M. and Hettne, B. (1985) 'The greening of the world: towards a non-deterministic model of global processes', pp. 204-70 in H. Addo *et al. Development as Social Transformation: reflections on the global problematique*, Hodder and Stoughton, Sevenoaks, for the United Nations University.

Friends of the Earth (1987) *Friends of the Earth Supporters' Newsletter* Winter: 3.

Frobel, F., Heinrichs, J. and Kreye, O. (1985) 'The global crisis and developing countries', pp. 111-24 in H. Addo *et al. Development as Social Transformation: reflections on the global problematique*. Hodder and Stoughton, Sevenoaks, for the United Nations University.

Frome, M. (1984) *Battle for the Wilderness*, Westview Press, Boulder, Colo.

Furon, R. (1947) *L'érosion du sol*, Payot, Paris.

Galois, B. (1976) 'Ideology and the idea of nature: the case of Peter Kropotkin', *Antipode* 8: 1-16.

Galtung, J. (1984) 'Perspectives on environmental politics in overdeveloped and underdeveloped countries', pp. 9-21 in B. Glaeser (ed.) *Ecodevelopment: concepts, projects, strategies*, Pergamon Press, Oxford.

Garcia-Lamor, J.-C. (ed.) (1984) *Public Participation in Development Planning*, Westview Press, Boulder, Colo.

George, C.J. (1973) 'The role of the Aswan High Dam in changing the fisheries of the southeastern Mediterranean', pp. 159-78 in M.T. Farvar and J.P. Milton (eds) *The Careless Technology: ecology and international development*, Stacey, London.

Ghatak, S. and Turner, R.K. (1978) 'Pesticide use in less developed countries: economic and environmental considerations', *Food Policy* 3: 136-46.

Gilbert, V.C. and Christy, E. J. (1981) 'The UNESCO Program on Man and the Biosphere (MAB)', pp. 701-20 in E.J. Kormondy and J.F. McCormick (eds) *Handbook of Contemporary Developments in World Ecology*, Greenwood Press, Westport, Conn.

Glaeser, B. (1984a) *Ecodevelopment in Tanzania: an empirical contribution on needs, self-sufficiency and environmentally sound agriculture on peasant farms*, Mouton, Berlin.

Glaeser, B. (ed.) (1984b) *Ecodevelopment: concepts, projects, strategies*, Pergamon Press, Oxford.

Glaeser, B. (ed.) (1987) *The Green Revolution Revisited: critique and alternatives*, Allen and Unwin, Hemel Hempstead.

Glaeser, B. and Vyasulu, V. (1984) 'The obsolescence of ecodevelopment?', pp. 23- 36 in B. Glaeser (ed.) *Ecodevelopment: concepts, projects, strategies*, Pergamon Press, Oxford.

Gliessman, S.R. (1984) 'Resource management in traditional agroecosystems: southeast Mexico', pp. 191-201 in G.K. Douglass (ed.) *Agricultural Sustainability in a Changing World Order*, Westview Press, Boulder, Colo.

Godana, B A (1985) *Africa's Shared Water Resources: legal and institutional aspects of the Nile, Niger and Senegal River systems*, Rienner, Boulder, Colo.

Goddman, C.R. (1976) 'Ecological aspects of water impoundment in the Tropics', *Revista de Biologia Tropical* 24 (Supplement 1) 87-112.

Goldsmith, E. (1987) 'Open letter to Mr. Conable, President of the World Bank', Ecologist 17: 58-61.

Goldsmith, E., Allen, R., Allaby, M., Davoll, J. and Lawrence, S. (1972) Blueprint for Survival, *The Ecologist* 2: 1-50 (also 1972, Penguin Books, Harmondsworth).

Goldsmith, E. and Hildyard, N. (1984) *Social and Environmental Impacts of Large Dams, Volume I*, Ecologist Magazine, Wadebridge, Cornwall.

Golub, R. and Townsend, J. (1977) 'Malthus, multinationals and the Club of Rome', *Social Studies of Science* 7: 202-22

Gonzales, N. 1972 'Sociology of a dam', *Human Organisation* 31: 353-60.

Good, K. (1986) 'The reproduction of weakness in the state and agriculture: the Zambian experience', *African Affairs* 85: 239-65.

Goodland, R.J. (1975) 'The tropical origins of ecology: Eugen Warming's Jubilee', *Oikos* 26: 240-5.

Goodland, R.J. (1978) *Environmental Assessment of the Tucurui Hydroproject, Rio Tocatins, Amazonia, Brazil*, Eletronorte SA, Brazilia.

Goodland, R.J. (1980a) 'Environmental rankings of Amazonian development projects in Brazil', *Environmental Conservation* 7: 9-26.

Goodland, R.J. (1980b) 'Indonesia's environmental progress in economic development', pp. 214-76 in V.H. Sutlive, N. Altshuler and M.D. Zamora (eds) *Where Have All The Flowers Gone? Deforestation in the Third World*, Studies in Third World Societies No. 13, College of William and Mary, Williamsburg, Va.

Goodland, R.J. (1984) *Environmental Requirements of the World Bank*, Environment and Science Unit, Projects Policy Department, World Bank, Washington, DC.

Goodland, R.J. and Irwin, H.S. (1975) *Amazon Jungle: green hell to red desert?*, Elsevier, Amsterdam.

Goodland, R.J. and Ledec, G. (1984) *Neoclassical Economics and Principles of Sustainable Development*, World Bank Office of Environmental Affairs, Washington, DC.

Gordon, R.J. (1985) 'Conserving bushmen to extinction in southern Africa', pp. 28-42 in *An End to Laughter? Tribal peoples and economic development*, Review No. 44, Survival International, London.

Gornitz, V. and NASA (1985) 'A survey of anthropogenic vegetation change in West Africa during the last century: climatic implications', *Climatic Change* 7: 285-326.

Gosling, L. (1979a) 'Resettlement losses and compensation', pp. 119-31 in Gosling L. (ed) *Population Resettlement in the Mekong River Basin*, Studies in Geography No. 10, University of North Carolina Press, Chapel Hill, NC.

Gosling, L. (ed) (1979b) *Population Resettlement in the Mekong River Basin*, Studies in Geography No. 10, University of North Carolina Press, Chapel Hill, NC.

Goudie, A.S. (1983a) 'The arid earth', pp. 152-71 in R. Gardner and H. Scoging (eds) *Mega-geomorphology*, Clarendon Press, Oxford.

Goudie, A.S. (1983b) 'Dust storms in space and time', *Progress in Physical Geography* 7: 503-30.

Goulet, D. (1971) *The Cruel Choice: a new concept in the theory of development*, Athanaeum, London.

Graf, W.L. (1980) 'The effect of dam closure on downstream rapids', *Water Resources Research* 16: 129-36.

Gradwohl, J. and Greenberg, R. (1988) *Saving the Tropical Forests*, Earthscan, London.

Graham, A. (1973) *The Gardeners of Eden*, Allen and Unwin, Hemel Hempstead.

Grainger, A. (1982) *Desertification: how people make deserts, how people can stop and why they don't*, Earthscan/IIED, London.

Green, K.M. (1983) 'Using Landsat to monitor tropical forest ecosystems', pp. 397-409 in S.L. Sutton, T.C. Whitmore and A.C. Chadwick (eds) *Tropical Rain Forest: ecology and management*, Blackwell, Oxford.

Greener, L. (1962) *High Dam Over Nubia*, Cassell, London.

Gregory, D. (1980) 'The ideology of control: systems theory and geography', *Tijdshrift voor Economische en Sociale Geographie* 81: 327-42.

Gregory, K.J. and Park, C.C. (1974) 'Adjustment of river channel capacity downstream from a reservoir', *Water Resources Research* 10: 870-7.

Grove, A.T. (1958) 'The ancient erg of Hausaland, and similar formations on the South side of the Sahara, *Geographical Journal* 124: 526-33.

Grove, A.T. (1973) 'A note on the remarkably low rainfall of the Sudan Zone in 1913', *Savanna* 2: 133-8.

Grove, A.T. (1977) 'Desertification', *Progress in Physical Geography* 1: 296-310.

Grove, A.T. (1981) 'The climate of the Sahara in the period of meteorological records', in J.A. Allan *Sahara: ecological change and early economic history*, Menas Press, London.

Grove, A.T. and Warren, A. (1968) 'Quaternary landforms and climate on the South side of the Sahara', *Geographical Journal* 134: 194-208.

Grove, R.H. (1987) 'Early themes in African conservation: the Cape in the Nineteenth Century', pp. 21-40 in D.M. Anderson and R.H. Grove (eds) *Conservation in Africa: people, policies and practice*, Cambridge University Press, Cambridge.

Guy, P.R. (1981) 'River bank erosion in the mid-Zambezi Valley downstream of Lake Kariba', *Biological Conservation* 19: 199-212.

Gwynne, M.D. (1982) 'The Global Environment Monitoring System (GEMS) of UNEP', *Environmental Conservation* 9: 35-42.

Hafner, J. A. (1979) 'Resettlement experience in replacement towns', pp. 85-106 in L. Gosling (ed) *Population resettlement in the Mekong River Basin*, Studies in Geography No. 10, University of North Carolina Press, Chapel Hill, NC.

Hailey, Lord (1938) *An African Survey*, Royal Institute for African Affairs and Oxford University Press, London.

Hamilton, A.C. (1982) *Environmental History of East Africa: a study of the Quaternary*, Academic Press, London.

Hamilton, A.C. (1984) *Deforestation in Uganda.*, Oxford University Press, Nairobi.

Hammer, D. (1976) 'EEC guidelines for Environmental Impact Assessment', pp. 35-44 in T. O'Riordan and R. Hey (eds) *Environmental Impact Assessment*, Saxon House, Farnborough.

Hamnet, I. (1970) 'A social scientist among technicians', *IDS Bulletin* 3: 24-9.

Handlos, W.L. and Williams, G.J. (1984) *Development of the Kafue Flats: the last five years*, Kafue Basin Research Committee, University of Zambia, Lusaka.

Hardin, G. (1974) 'Living on a lifeboat', *Bioscience* 24: 561-8.

Hardonjo, J.M. (1977) *Transmigration in Indonesia*, Oxford University Press, Selangor, Malaysia.

Hardonjo, J.M (1983) 'Rural development in Indonesia: the 'top down' approach', pp. 38-66 in D.A.M. Lea and D.P. Chaudhri (eds) *Rural Development and the State: contradictions and dilemmas in developing countries*, Methuen, London.

Hare, F.K. (1984) 'Recent climatic experience in the arid and semi-arid lands', *Desertification Control Bulletin* 10: 15-20

Harlan, J.R. and Pasquerau, J. (1969) 'Décrue agriculture in Mali', *Economic Botany* 23: 70-4.

Harrell-Bond, B. (1985) 'Humanitarianism in a straightjacket', *African Affairs* 84: 3-14.

Harrison, P. (1987) *The Greening of Africa: breaking through in the battle for land and food*, Paladin, London.

Harrison, P. and Palmer, R. (1986) *News out of Africa: Biafra to Band Aid*, Hilary Shipman, London.

Harrison, T.H. (1933) 'The Oxford University Expedition to Sarawak, 1932', *Geographical Journal* 82: 385-410.

Harroy, J.-P. (1949) *Afrique: terre qui meurt: la dégradation des sols Africains sous l'influence de la colonisation*, Marcel Hayez, Brussels.

Hart, D. (1980) *The Volta River Project: a case study in politics and technology*, Edinburgh University Press, Edinburgh.

Hart, K. (1982) *The Political Economy of West African Agriculture*, Cambridge University Press, Cambridge.

Harvey, D. (1974) 'Population, resources and the ideology of science', *Economic Geography* 50: 256-77.

Haugerud, A. (1986) 'An anthropologist in an African Research Institute: an informal essay', *Development Anthropology Network* 4: 4-9.

Hay, P.R. and Hayward, M.G. (1988) 'Comparitive green politics: beyond the European context?', *Political Studies* 36: 433-48.

Hayes, F.R. (1953) 'Artificial freshets and other factors controlling the ascent of a population of Atlantic salmon in the Lettavre River, Nova Scotia', *Bulletin of the Fisheries Resources Board of Canada* 99: 1-47.

Hays, S.P. (1959) *Conservation and the Gospel of Efficiency: the progressive conservation movement 1890-1920*, Harvard University Press, Cambridge, Mass.

Hays, S.P. (1987) *Beauty, Health and Permanence: environmental politics in the United States (1955-1985)*, Cambridge University Press, Cambridge.

Hecht, S.B. (1980) 'Deforestation in the Amazon basin: magnitude, dynamics and soil resource effects', pp. 61-108 in V.H. Sutlive, N. Altshuler and M.D. Zamora (eds) *Where Have All The Flowers Gone? Deforestation in the Third World*, Studies in Third World Societies No. 13, College of William and Mary, Williamsburg, Va.

Hecht, S.B. (1984) 'Cattle ranching in Amazonia: political and ecological considerations', pp. 366-400 in M. Schmink and C.H. Wood (eds) (1984) *Frontier Expansion in Amazonia*, University of Florida Press, Gainesville, Fla.

Hecht, S.B. (1985) 'Environment, development and politics: capital accumulation and the livestock sector in Eastern Amazonia', *World Development* 13: 663-84.

Hemming, J. (1985) *Change in the Amazon River Basin*, (2 volumes). Manchester University Press, Manchester.

Henderson-Sellers, A. and Gornitz, V. (1984) 'Possible climatic impacts of land cover transformations, with particular emphasis on Tropical deforestation', *Climatic Change* 6: 231-58.

Herlocker, D. (1979) *Vegetation of Southwestern Marsabit District Kenya*, Integrated Project on Arid Lands Technical Report No. D-1 (Man and the Biosphere Project 3: Impact of Human Land Use Practices on Grazing Lands), Integrated Project on Arid Lands, Nairobi.

Herrera, R., Jordan, C.F., Medina, E. and Klinge, H. (1981) 'How human activities disturb the nutrient cycle of a tropical rainforest in Amazonia', *Ambio* 10: 109-14.

Hickling, C.F. (1961) *Tropical Inland Fisheries*. Longmans, London.

Higgins, G.M., Kassam, A.H. and Naiken, L. (1982) *Potential Population Supporting Capacities of Lands in the Developing World*, Food and Agriculture Organisation, Rome.

Hildermeier, M. (1979) 'Agrarian social protest, populism and economic development: some problems and results from recent studies', *Social History* 4: 319-32.

Hill, A.W. (1925) 'The best means of promoting a complete botanical survey of the different parts of the empire', pp. 196-204 in F.T. Brooks (ed.) *Imperial Botanical Conference, London 1924: report of proceedings*, Cambridge University Press, Cambridge.

Hill, F. (1978) 'Experiments with a public sector peasantry: agricultural schemes and class formation in Africa', pp. 25-41 in A.K. Smith and C.E. Welch (eds) *Peasants in Africa*. African Studies Association, Boston, Mass.

Hill, P. (1977) *Population, Prosperity and Poverty: rural Kano 1900 and 1970*, Cambridge University Press, Cambridge.

Hill, P. (1986) *Development Economics On Trial: the anthropological case for the prosecution*, Cambridge University Press, Cambridge.

Hilton, T.E. and Kuwo-Tsri, J.Y. (1970) 'The impact of the Volta Scheme on the Lower Volta floodplains', *Journal of Tropical Geography*. 30: 29-37.

Hingston, R.W.G. (1930) 'The Oxford University expedition to British Guiana', *Geographical Journal* 76: 1-24.

Hingston, R.W.G. (1931) 'Proposed British National Parks for Africa', *Geographical Journal* 77: 401-28.

Hiraoka, M. and Yamamoto, S. (1980) 'Agricultural development in the Upper Amazon of Ecuador', *Geographical Review* 70: 423-45.

Hobley, C.W. (1914) 'The alleged desiccation of East Africa', *Geographical Journal* 44: 467-77.

Hogg, R. (1983) 'Irrigation agriculture and pastoral development: a lesson from Kenya', *Development and Change* 14: 577-91.

Hogg, R. (1987a) 'Development in Kenya: drought, desertification and food security', *African Affairs* 86: 47-58.

Hogg, R. (1987b) 'Settlement, pastoralism and the commons: the ideology and practice of irrigation development in northern Kenya', pp. 293-306 in D.M. Anderson and R.H. Grove (eds) *Conservation in Africa: people, policies and practice*, Cambridge University Press, Cambridge.

Holden, C. (1987) 'World Bank launches new environmental policy', *Science* 236: 769.
Holden, C. (1988) 'The greening of the World Bank', *Science* 240: 1610-11.
Holdgate, M.W., Kassas, M. and White, G.F. (1982a) 'World environmental trends between 1972 and 1982', *Environmental Conservation* 9: 11-29.
Holdgate, M.W., Kassas, M. and White, G.F. (eds) (1982b) *The World Environment 1972-198,* A report by the United Nations Environment Programme, Tycooly International, Dublin.
Hollis, G.E. (1978) 'The falling levels of the Caspian and Aral Seas', *Geographical Journal* 144: 62-80.
Homewood, K. and Rodgers, W.A. (1984) 'Pastoralism and conservation', *Human Ecology* 12: 431-41 .
Homewood, K. and Rodgers, W.A. (1987) 'Pastoralism, conservation and the overgrazing controversy', pp. 111-28 in D.M. Anderson and R.H. Grove (eds) *Conservation in Africa: people, policies and practice,* Cambridge University Press, Cambridge.
Hopper, W. (1988) *The World Bank's Challenge: balancing economic need with environmental protection,* 7th Annual World Conservation Lecture, 3 March, World Wide Fund for Nature UK, Godalming.
Horberry, J.A.J. (1988) 'Fitting USAID to the environmental assessment provisions of NEPA', pp. 286-99 in P. Wathern (ed) *Environmental Impact Assessment: theory and practice,* Unwin Hyman, London.
Horowitz, M.M. (1987) 'Destructive development', *Institute of Development Anthropology Network 5*: 1-2.
Horowitz, M.M. and Little, P.D. (1987) 'African pastoralism and poverty: some implications for drought and famine', pp. 59-82 in M. Glantz (ed.) *Drought and Hunger in Africa: denying famine a future,* Cambridge University Press, Cambridge.
Howard, G.W. and Williams, G.J. (eds) (1982) *The Consequences of Hydroelectric Power Development on the Kafue Flats,* Kafue Basin Research Project, Lusaka (Proceedings of the National Seminar on Environment and Change, Lusaka).
Howarth, D. (1961) *The Shadow of the Dam,* Collins, London.
Howell, P.P. (1953) 'The Equatorial Nile Project and its effects in the Sudan, *Geographical Journal* 119: 33-48 .
Howell, P.P. (1983) 'The impact of the Jonglei Canal in the Sudan', *Geographical Journal* 149: 286-300.
Howell, P., Lock, M. and Cobb, S. (1988) *The Jonglei Canal: impact and opportunity,* Cambridge University Press, Cambridge.
Hughes, F.M.R. (1983) 'Helping the World Conservation Strategy? Aid agencies on the Tana River in Kenya', *Area* 15: 177-83.
Hughes, F.M.R. (1984) 'A comment on the impact of development schemes on the floodplain forests of the Tana River of Kenya', *Geographical Journal* 150: 230-44.
Hughes, F.M.R. (1985) 'The impact of development schemes on the Tana River floodplain, Kenya', unpublished Ph.D. thesis, University of Cambridge.
Hughes, F.M.R. (1987) 'Conflicting uses for forest resources in the lower Tana River Basin of Kenya', pp. 211-28 in D.M. Anderson and R.H. Grove (eds)

Conservation in Africa: people, policies and practice, Cambridge University Press, Cambridge.
Hughes, F.M.R. (1988) 'The ecology of African floodplain forests in semi-arid zones: a review', *Journal of Biogeography* 15: 127-40.
Hulme, M. (1987) 'Secular changes in wet season structure in central Sudan', *Journal of Arid Environments* 13: 31-46.
Huxley, J.L. (1930) *African View*, Chatto and Windus, London.
Huxley, J.L. (1977) *Memories II*, Harper and Row, New York.
Hyden, G. (1980) *Beyond Ujamaa in Tanzania: underdevelopment and an uncaptured peasantry*, Heinemann, London.
Hynes, H.B.N. (1970) *The Ecology of Running Waters*, University of Toronto Press, Toronto.
ICOLD (1980) *Dams and their Environment*, International Commission on Large Dams, Paris.
ICOLD (1981) *Dam Projects and Environmental Success*, International Commission on Large Dams, Paris.
Idso, S.B. (1977) 'A note on some recently proposed mechanisms of genesis of deserts', *Quarterly Journal of the Royal Meteorological Society* 103: 369-70.
IIkporukpo, C.O. (1983) 'Environmental deterioration and public policy in Nigeria', *Applied Geography* 3: 303-16.
Illich, I. (1973) 'Outwitting the "developed" countries', pp. 357-68 in H. Bernstein (ed.) *Underdevelopment and Development: the Third World today*, Penguin, Harmondsworth.
IUCN (1980) *The World Conservation Strategy*, International Union for Conservation of Nature and Natural Resources, United Nations Environment Programme, World Wildlife Fund, Geneva.
IUCN (1984a) *National Conservation Strategies: a framework for sustainable development*, IUCN, Geneva.
IUCN (1984b) *Towards Sustainable Development: a national conservation strategy for Zambia*, IUCN, Geneva.
IUCN Commission on National Parks and Protected Areas (1984) 'Categories, objectives and criteria for protected areas', pp. 47-53 in J.A. McNeely and K.R. Miller (eds) *National Parks, Conservation and Development: the role of protected areas in sustaining society*, Smithsonian Institute Press, Washington, DC.
Jacks, G.V. and Whyte, R.O. (1938) *The Rape of the Earth: a world survey of soil erosion*, Faber and Faber, London.
Jackson, P.B.N. (1966) 'The establishment of fisheries in man-made lakes in the tropics', in R.H. Lowe-McConnell (ed.) *Man-made Lakes*, Institute of Biology and Academic Press, London.
Jackson, R.D. and Idso, S.B. (1975) 'Surface albedo and desertification', *Science* 189: 1012-13.
Jago, N. (1987) 'The return of the eighth plague', *New Scientist* 18 June: 47-51.
Jodha, N.S. (1980) 'Intercropping in traditional farming systems', *Journal of Development Studies* 16: 427-42.
Johns, A.D. (1985) 'Selective logging and wildlife conservation in tropical rain-forest: problems and recommendations', *Biological Conservation* 31: 355-76.

Johnson, B. (1985) 'Chimera or opportunity? An environmental appraisal of the recently concluded International Tropical Timber Agreement', *Ambio* 14: 42-4.

Johnson, B. and Blake, R.O. (1980) *The Environment and Bilateral Aid Agencies,* IIED, Washington, DC.

Jones, B. (1938) 'Desiccation and the West African Colonies', *Geographical Journal* 41: 401-23.

Jones, G.H. (1936) *The Earth Goddess: a study of native farming in the West African context,* Longman, Green and Co., London.

Jubb, R.A. (1972) 'The J.G. Strydon Dam, Pongolo River, northern Zululand: the importance of floodplain pans below it', *Piscator* 86: 104-9.

Justice, C.O., Townshend, J.R.G., Holben, B.N. and Tucker, C.J. (1985) 'Analysis of the phenology of global vegetation using meteorological satellite data', *International Journal of Remote Sensing* 6: 1271-318.

Kabuzya, B.J., Mtalo, F.W. and Tesako, E. (1980) 'Hydraulic studies in lower Rufiji River, Tanzania, for consequence of dam project at Steigler's Gorge', pp. 36-42 in S. Bruk (ed.) *Hydraulic Research and River Basin Development in Africa,* Proceedings of the IAHR/UNESCO Seminar, Nairobi, September, International Association of Hydrological Research.

Kahn, H. (1979) *World Economic Development: 1979 and beyond,* Croom Helm, London.

Kahn, H. and Wiener, A.J. (1967) *The Year 2000: a framework for speculation on the next 33 years,* Macmillan, London.

Karrar, G. (1984) 'The UN Plan of Action to Combat Desertification and the concommitent UNEP campaign', *Environmental Conservation* 11: 99-102.

Kartawinata, K., Adisoemarto, S., Riswar, S. and Vayda, A.P. (1980) 'The environmental consequences of the removal of the forest in Indonesia', pp. 191-214 in V.H. Sutlive, N. Altshuler, M.D. Zamora, (eds) *Where Have All The Flowers Gone? Deforestation in the Third World,* Studies in Third World Societies No. 13, College of William and Mary, Williamsburg, Va.

Kartawinata, K., Adisoemarto, S., Riswa, S. and Vayda, A.P. (1981) 'The impact of man on a tropical forest in Indonesia', *Ambio* 10: 115-19.

Kassam, A.H. and Stockinger, K.R. (1973) 'Growth and nitrogen uptake of sorghum and millet in mixed cropping', *Samaru Agricultural Newsletter* 15: 28-32.

Kassas, M. (1973) 'Impact of river control schemes on the shoreline of the Nile Delta', pp. 179-88 in M.T. Farvar and J.P. Milton (eds) *The Careless Technology: ecology and international development,* Stacey, London.

Kellerhals, R. (1982) 'Effect of river regulation on channel stability', pp. 685-716 in R.D. Hey, J.C. Bathurst and C.R. Thorne (eds) *Gravel Bed Rivers,* Wiley, Chichester.

Kemf, E. (1987) 'WWF seeks alternatives to massive pesticide plan', *WWF News* September/October 49: 1,5.

Kennedy, W.V. (1988) 'Environmental Impact Assessment and bilateral development aid: an overview', pp. 272-82 in P. Wathern (ed.) *Environmental Impact Assessment: theory and practice,* Unwyn Hyman, London.

Kerssens, P.J.M. and Zwaard, J.J. van der (1980) 'Hydraulic and morphological impacts of dam construction', pp. 43-50 in S. Bruk (ed.) *Hydraulic Research and River Basin Development in Africa,* Proceedings of the IAHR/UNESCO Seminar, Nairobi, September, International Association of Hydrological Research.

Khogali, M.M. (1982) 'The problem of siltation in Khasm el Girba Reservoir: its implications and suggested solutions', pp. 96-106 in H.G. Mensching (ed.) *Problems of the Management of Irrigated Land in Areas of Traditional and Modern Cultivation,* International Geographical Union Working Group on Resource Management in Drylands, Hamburg.

King, J.A. (1967) *Economic Development Projects and their Appraisal: cases and principles from the experience of the World Bank,* Johns Hopkins University Press, Baltimore, Md.

Kirby, J.P. (1987) 'Why the Anufo do not eat frogmeat: the implications of taboo-making for development work', *African Affairs* 86: 59-72.

Kirchner, J.W., Ledec, G., Goodland, R.J.A and Drake, J.M. (1985) 'Carrying capacity, population growth and sustainable development' , pp. 42-89 in D.J. Mahar (ed.) *Rapid Population Growth and Human Carrying Capacity,* World Bank Staff Working Paper No. 690, Washington, DC.

Kishk, M.A. (1986) 'Land degradation in the Nile Valley', *Ambio* 15: 226-30.

Kitching, G. (1982) *Development and Underdevelopment in Historical Perspective: populism, nationalism and industrialism,* Methuen, London.

Klein, M. (ed.) (1980) *Peasants in Africa: historical and contemporary perspectives,* Sage, Beverly Hills and London, .

Komarov, B. (1978) *The Destruction of Nature in the Soviet Union,* Pluto Press, London.

Kormondy, E.J. and McCormick, J.F. (eds) (1981) *Handbook of Contemporary World Developments in Ecology,* Greenwood Press, Westport, Conn.

Kowal, J.M. and Adeoye, K.B. (1973) 'An assessment of aridity and the severity of the 1972 drought in northern Nigeria and neighbouring countries', *Savanna* 2: 145-58.

Kowal, J.M. and Kassam, A.H. (1978) *Agricultural Ecology of Savanna: a study of West Africa,* Clarendon Press, Oxford.

Kropotkin, P. (1972) *The Conquest of Bread,* Allen Lane, London (edited by P. Avrich; first published in English as articles in *Freedom,* 1892-4, and as a book in 1906, London)

Kropotkin, P. (1974) *Fields, Factories and Workshops Tomorrow,* Allen and Unwin, London (edited by C. Ward; first edition Hutchinson, London, 1899).

Lagler, K.F. (ed.) (1969) *Man-made Lakes: planning and development,* Food and Agriculture Organisation, Rome.

Lagler, K.F. (1971) 'Ecological effects of hydrological dams', pp. 133-57 in D.A. Berkowitz and A.M. Squires (eds) *Power Generation and Environmental Change,* MIT Press, Cambridge, Mass.

Lako, G.T. (1985) 'The impact of the Jonglei Scheme on the economy of the Dinka', *African Affairs* 84: 15-38.

Lal, R. (1981) 'Deforestation of tropical rainforest and hydrological problems', pp. 130-40 in R. Lal. and E.W. Russell (eds) *Tropical Agricultural Hydrology: watershed management and land use,* Wiley, Chichester.

Lamb, D. (1980) 'Some ecological consequences of logging rainforests on Papua New Guinea', pp. 55-64 in J.I. Furtado (ed.) *Tropical Ecology and*

Development, International Society for Tropical Ecology, Kuala Lumpur (Proceedings of the 5th International Symposium on Tropical Ecology).

Lamb, P.J. (1979) 'Some perspectives on climate and climatic dynamics', *Progress in Physical Geography* 3: 215-35.

Lamb, P.J. (1982) 'Persistence of subsaharan drought', *Nature* 299: 46-7.

Lanly, J.-P. (1982) *Tropical Forest Resources*, Forestry Paper No. 30, Food and Agriculture Organisation, Rome.

Lanning, G. and Mueller, M. (1979) *Africa Undermined: mining companies and the underdevelopment of Africa*, Penguin, Harmondsworth.

Lawrence, P. (ed.) (1986) *World Recession and the Food Crisis in Africa*, Review of African Political Economy and James Currey, London.

Lawson, R.M. (1963) 'The economic organisation of the Egeria fishing on the River Volta', *Proceedings of the Malacological Society of London* 35: 273-87.

Lea, D.A.M. and Chaudhri, D. P. (1983) 'The nature, problems and approaches to rural development', pp. 1-37 in D.A.M. Lea and D.P. Chaudhri (eds) *Rural Development and the State: contradictions and dilemmas in developing countries*, Methuen, London.

Lee, E. (1981) 'Basic needs strategies: a frustrated response to development from below?', pp. 107-27 in W.B. Stohr and D.R.F. Taylor (eds) *Development: from above or below?*, Wiley, Chichester.

Lee, N. (1982) 'The future development of Environmental Impact Assessment', *Journal of Environmental Management* 14: 71-90.

Lee, N. (1983) 'Environmental Impact Assessment: a review', *Applied Geography* 3: 5-28.

Lehmann, D. (1986) *Dependencia: an ideological history*, Institute of Development Studies Discussion Paper 219, IDS, Brighton.

Lelek, A. and El-Zarka, S. (1973) 'Kainji Lake, Nigeria, pp. 655-60 in W.C. Ackerman, G.F. White and E.B. Worthington (eds) *Man-made Lakes: their problems and environmental effects*, Geophysical Monograph 17, American Geophysical Union, Washington, DC.

Lenin, V.I. (1972) *Imperialism, the Highest Stage of Capitalism*, Progess Publishers, Moscow.

Leopold, L.B., Clarke, F.E., Nanshaw, B.B. and Balsley, J.R. (1971) *A Procedure for Evaluating Environmental Impact*, US Geological Survey Circular 645, Washington, DC.

Leys, C. (1975) *Underdevelopment in Kenya: the political economy of neo-colonialism*, Heinemann, London.

Lightfoot, R.D. (1978) 'The costs of resettling reservoir evacuees in NE Thailand', *Journal of Tropical Geography* 47: 63-74.

Lightfoot, R.P. (1979) 'Alternative resettlement strategies in Thailand: lessons from experience', pp. 28-38 in L. Gosling (ed.) *Population Resettlement in the Mekong River Basin*, Studies in Geography No. 10, University of North Carolina Press, Chapel Hill, NC.

Lightfoot, R.P. (1981) 'Problems of resettlement in the development of river basin in Thailand', pp. 93-114 in S.K. Saha and C.J. Barrow (eds) *River Basin Planning: theory and practice*, Wiley, Chichester.

Linares, O.F. (1981) 'From tidal swamp to inland valley: on the social organisation of wet rice cultivation among the Diola of Senegal', *Africa* 51: 557-95.

Lindsay, W.K. (1987) 'Integrating parks and pastoralists: some lessons from Amboseli', pp. 149-68 in D.M. Anderson and R.H. Grove (eds) *Conservation in Africa: people, policies and practice*, Cambridge University Press, Cambridge.

Linear, M. (1981) 'Zapping Africa's flies', *Vole* March: 14-18.

Linney, W. and Harrison, S. (1981) *Large Dams and the Developing World: social and ecological costs and benefits - a look at Africa*, Environment Liason Centre, Nairobi.

Lipton, M. (1977) *Why Poor People Stay Poor: a study of urban bias in world development*, Temple Smith, London.

Little, T. (1965) *High Dam at Aswan: the subjugation of the Nile*, John Day, New York.

Liu, Kam-biu (1985) 'Forest changes in the Amazon Basin during the last glacial maximum', *Nature* 318: 556-7.

Lock, M.A. and Williams, D.D. (eds) (1981) *Perspectives in Running Water Ecology*, Plenum, New York.

Lockwood, J. (1986) 'The causes of drought with particular reference to the Sahel', *Progress in Physical Geography* 10: 111-19.

Lohani, B.N. (1982) 'Establishing water quality objectives and emission targets', *Water Supply and Management* 6: 469-85.

Lohani, B.N. and Thanh, N.C. (1980) 'Impacts of rural development and their assessment in southeastern Asia', *Environmental Conservation* 7: 213-18.

Lovejoy, T.E., Bierregaard, R.O., Rankin, J. and Schubart, H.O.R. (1983) 'Dynamics of forest fragments', pp. 377-84 in S.L. Sutton, T.C. Whitmore and A.C. Chadwick (eds) *Tropical Rain Forest: ecology and management*, Blackwell, Oxford.

Low, D.A. and Lonsdale, J.M. (1976) 'Introduction: towards a new order 1945-1963', pp. 1-63 in D.A. Low and A. Smith (eds) *History of East Africa, Volume III*, Clarendon Press, Oxford.

Lowe, P.D. (1976) 'Amateurs and professionals: the institutional emergence of British plant ecology', *Journal of the Society for the Bibliography of Natural History* 7: 517-35.

Lowe, P.D. (1983) 'Values and institutions in the history of British nature conservation' pp. 329-52 in A. Warren and F.B. Goldsmith (eds) *Conservation in Perspective*, Wiley, Chichester.

Lowe, P.D. and Goyder, P. (1983) *Environmental Groups in Politics*, Allen and Unwin, Hemel Hempstead.

Lowe, P.D. and Warboys, M. (1975) 'Ecology and the end of ideology', *Antipode* 10: 12-21.

Lowe, P.D. and Warboys, M. (1976) 'The ecology of ecology', *Nature* 262: 432-3.

Lowe-McConnell, R.H. (ed.) (1966) *Man-made Lakes*, Institute of Biology and Academic Press, London.

Lowe-McConnell, R.H. (1975) *Fish Communities of Tropical Freshwaters*, Longman, London.

Lowe-McConnell, R.H. (1985) 'The biology of the river systems with particular reference to the fish', pp. 101-40 in A.T. Grove (ed.) *The Niger and its Neighbours: environment, history and hydrobiology, human use and health hazards of the major West African rivers,* Balkema, Rotterdam.

Lugard, Lord (1922) *The Dual Mandate in British Tropical Africa*, Frank Cass, London (1965 edition).

Lumsden, D.P. (1975) 'Towards a systems model of stress: feedback from an anthropological study of the impact of Ghana's Volta River Project', *Stress and Anxiety* 2: 191-227.

Mabbutt, J.A. (1984) 'A new global assessment of the status and trends of desertification', *Environmental Conservation* 11: 103-13.

Mabogunje, A.L. (ed.) (1973) *Kainji: a Nigerian man-made lake,* Kainji Lake Studies Volume 2: *Socio-economic conditions*, Nigerian Institute for Social and Economic Research, Ibadan.

Mabogunje, A.L. (1980) *The Development Process: a spatial perspective,* Hutchinson, London.

McCormick, J.S. (1985) *Acid Earth: the global threat of acid pollution,* Earthscan, London.

McCormick, J.S. (1986a) 'Britain, the United States and the international environmental movement 1945-1982', unpublished M.Phil. thesis, University College London.

McCormick, J.S. (1986b) The origins of the World Conservation Strategy. *Environmental Review* 10 (2): 177-87.

McCormick, J.S. (1989) *Reclaiming Paradise: the global environmental movement*, Indiana University Press, Bloomington, Ind.

McHale, J. and McHale, M.C. (1978) *Basic Human Needs: a framework for action*, Transaction Books, New Brunswick, NJ.

McIntosh, R.P. (1985) *The Background of Ecology: concept and theory,* Cambridge University Press, Cambridge.

MacKenzie, D. (1987) 'Thousands poisoned by pesticide in Guyana', *New Scientist* 19 March: 18

Mackenzie, J.M. (1987) 'Chivalry, social Darwinism and ritualised killing: the hunting ethos in Central Africa up to 1914', pp. 41-62 in D.M. Anderson and R.H. Grove (eds) *Conservation in Africa: people, policies and practice,* Cambridge University Press, Cambridge.

McLean, R.C. (1919) 'Studies in the ecology of Tropical Rain Forest: with special reference to the forests of South Brazil', *Journal of Ecology* 7: 121-72.

McNeely, J.A. (1984) 'Introduction: protected areas are adapting to new realities', pp. 1-7 in J.A. McNeely and K.R. Miller (eds) *National Parks, Conservation and Development: the role of protected areas in sustaining society*, Smithsonian Institute Press, Washington, DC.

McNeely, J.A. and Miller, K.R. (eds) (1984) *National Parks, Conservation and Development: the role of protected areas in sustaining society*. Smithsonian Institute Press, Washington, DC.

McNeely, J.A. and Pitt, D. (eds) (1987) *Culture and Conservation: the human dimension in environmental planning*, Croom Helm, London.

McRobie, G. (1981) *Small Is Possible*, Jonathan Cape, London.

Maddox, J. (1972) *The Doomsday Syndrome*, MacMillan, London.

Mann, D.W. (1981) 'Fewer people for a better world: a plea for negative population growth', *Environmental Conservation* 8: 260-1.

Margalef, R. (1968) *Perspectives in Ecological Theory*, University of Chicago Press, Chicago.

Marsh, G.P. (1864) *Man and Nature; or, physical geography as modified by human action*, Scribners, New York; republished in 1965 by Belknap Press, Cambridge, Mass., edited by D. Lowenthal.

Matthieson, P. and Douthwaite, B. (1985) 'The impact of tsetse control campaigns on African wildlife', *Oryx* 19: 202-9.

May, B. (1981) *The Third World Calamity*, Routledge and Kegan Paul, London.

Maybury-Lewis, D. (1984) 'Demystifying the second conquest', pp. 127-34 in M. Schmink and C.H. Wood, (eds) (1984) *Frontier Expansion in Amazonia*, University of Florida Press, Gainesville, Fla.

Meadows, D., Randers, J. and Behrens, W.W. (1972) *The Limits to Growth*, Universe Books, New York.

Melillo, J.M., Palm, C.A., Houghton, R.A., Woodwell, G.M. and Myers, N. (1985) 'A comparison of two recent estimates of forest disturbance in tropical forests', *Environmental Conservation* 12: 37-40

Merchant, C. (1980) *The Death of Nature: women, ecology and the scientific revolution*, Harper and Row, New York.

Meyers, S. (1976) 'US experience with national environmental impact legislaton', pp. 45-55 in T. O'Riordan and R.D. Hey (eds) *Environmental Impact Assessment*, Saxon House, Farnborough.

Michel, P. (1980) 'The southwest Sahara margin: sediments and climatic changes during the recent Quaternary', *Palaeoecology of Africa* 12: 297-306.

Middleton, N.J. (1985) 'Effect of drought on dust production in the Sahel', *Nature* 316: 431-4

Mies, M. (1986) *Patriarchy and Accumulation on a World Scale: women in the international division of labour*, Zed Books, London.

Migdal, J. S. (1976) *Peasants, Politics and Revolution: pressures towards social change in the Third World*, Princeton University Press, Princeton, NJ.

Mishan, E.J. (1969) *The Costs of Economic Growth*, Penguin Books, Harmondsworth (first published Staples Press, 1967).

Mishan, E.J. (1977) *The Economic Growth Debate: an assessment*, Allen and Unwin, Hemel Hempstead.

Moghraby, A.I. el (1982) 'The Jonglei Canal: needed development or potential ecodisaster?', *Environmental Conservation* 9: 141-8.

Moghraby, A.I. el and Sammani, M.O. el (1985) 'On the environmental and socio-economic impact of the Jonglei Canal Project, S. Sudan', *Environmental Conservation* 12: 41-8.

Mohun, J. and Sattaur, O. (1987) 'The drowning of a culture', *New Scientist* 15 January: 37-42.

Moore, N.W. (1987) *The Bird of Time*, Cambridge University Press, Cambridge.

Moran, E.F. (ed.) (1983) *The Dilemma of Amazonian Development*, Westview Press, Boulder, Colo.

Morgan, A.E. (1971) *Dams and Other Disasters: a century of the Army Corps of engineers in civil works*, Porter Sergent, Boston, Mass.

Moris, J. and Norman, R. (1984) *Prospects of Small Scale Irrigation in the Sahel*, USAID Water Management Synthesis II, Logan, Utah.
Morris, M.D. (1979) *Measuring the Condition of the World's Poor: the physical quality of life index*, Pergamon Press, Oxford.
Mosley, P. (1986) 'The politics of economic liberalisation: USAID and the World Bank in Kenya 1980-84', *African Affairs* 85: 107-19.
Mote, V.L. and Zumbrunne, C. (1977) 'Anthropogenic environmental alteration of the Sea of Azov', *Soviet Geography* 18: 744-59.
Munn, R.E. (ed.) (1979) *Environmental Impact Assessment: impacts and procedures*, SCOPE Report No. 5, Wiley, Chichester.
Munro, D.A. (1978) 'The thirty years of IUCN', *Nature and Resources* 14(2): 14-18.
Munslow, B., Katere, V., Ferf, A. and O'Keefe, P. (1988) *The Fuelwood Trap: a study of the SADCC region*, Earthscan, London.
Murdia, R. (1982) 'Forest development and tribal welfare: analysis of some policy issues', pp. 31-41 in E.G. Hallsworth (ed.) *Socio-economic Effects and Constraints in Tropical Forest Management*, Wiley, Chichester.
Murray, S.P., Coleman, J.M., Roberts, H.H. and Salama, M. (1981) 'Accelerated currents and sediment transport of the Damieth Nile promontory', *Nature* 293: 51-4.
Myers, N. (1980) 'Conversion of tropical moist forests', National Academy of Sciences, Washington, DC.
Myers, N. (1981) 'The hamburger connection: how Central America's forests become North America's hamburgers', *Ambio* 10: 3-8.
Myers, N. (1984) *The Primary Source: tropical forests and our future*, Norton, New York.
Myers, N. (ed.) (1985) *The Gaia Atlas of Planet Management*, Pan Books, London.
Myers, N. and Myers, D. (1982) 'From "duck pond" to the global commons: increasing awareness of the supranational nature of emerging environmental issues', *Ambio* 11: 195-201.
Narayanayaya, D.V. (1928) 'The aquatic weeds in Deccan irrigation canals', *Journal of Ecology* 16: 123-33.
Nash, R. (1973) *Wilderness and the American Mind*, Yale University Press, New Haven, Conn.
Nature (1948) 'Aspects of colonial development', *Nature* 162: 547-50.
Nature (1952) 'Development of backward countries', *Nature* 170: 677-80.
Nelson, J.G. (1987) 'National Parks and protected areas, National Conservation Strategies and sustainable development', *Geoforum* 18: 291-320.
New Left Review (1974) 'Themes', *New Left Review* 8: 1-2.
Nicholaides, J.J. II, Bandy, D.E., Sanchez, P.A., Villachica, J.H., Contu, A.J., and Valverde, C.S. (1984) 'Continuous cropping potential in the Upper Amazon Basin', pp. 337-65 in M. Schmink and C.H. Wood (eds.) *Frontier Expansion in Amazonia*, University of Florida Press, Gainesville, Fla.
Nicholson, E.M. (1975) 'Conservation', pp. 12-14 in E.B. Worthington (ed.) *The Evolution of the IBP*, Cambridge University Press, Cambridge.
Nicholson, S.E. (1978) 'Climatic variation in the Sahel and other African regions during the past five centuries', *Journal of Arid Environments* 1: 3-24.

Nicholson, S.E. (1980) 'Saharan climates in historic times', pp. 173-200 in M.A.J. Williams and H. Faure (eds) *The Sahara and the Nile: Quaternary environments and prehistoric occupation in northern Africa*, Balkema, Rotterdam.

Nicholson, S.E. (1988) 'Land surface atmosphere interaction: physical processes and surface changes and their impact', *Progress in Physical Geography* 12: 36-55.

Norton-Griffiths, M. (1979) 'The influence of grazing, browsing and fire on the vegetation dynamics of the Serengeti', pp. 310-52 in A.R.E. Sinclair and M. Norton-Griffiths (eds) *Serengeti: dynamics of an ecosystem*, University of Chicago Press, Chicago.

Nye, P. and Greenland, D. (1960) *The Soil Under Shifting Cultivation*, Technical Report 51, Commonwealth Agricultural Bureau, London.

Nyerere, J. K. (1985) 'Africa and the debt crisis', *African Affairs* 84: 489-98.

Obeng, L.E. (ed.) (1969) *International Symposium on Man-made Lakes, Accra*, Ghana University Press, Accra.

Oculi, O. (1981) 'Planning the Bakolori irrigation project', *Food Policy* 6: 201-4.

ODA (1987) *The Environment and the British Aid Programme*, Overseas Development Administration, HMSO, London.

Oglesby, R.T., Carlson, C.A. and McCann, J.A. (1972) *River Ecology and Man*, Academic Press, New York.

Ogoyuntibo, J.S. and Richards, P. (1977) 'The extent and intensity of the 1969-1973 drought in Nigeria: a provisional analysis', pp. 114-26 in D. Dalby, R.J. Harrison Church and F. Bezzaz (eds) *Drought in Africa 2*, International African Institute, London.

Ogoyuntibo, J.S. and Richards, P. (1978) 'Drought and the Nigerian farmer', *Journal of Arid Environments* 1: 169-94.

Oladimeji, A. and Wade, A.W. (1984) 'Effects of effluents from a sewage treatment plant on the aquatic organisms', *Water Air and Soil Pollution* 23: 309-16.

Oldfield, S. (1988) 'Rare tropical timbers', IUCN, Gland, Switzerland.

Olofin, E. (1984) 'Some effects of the Tiga Dam on valleyside erosion in downstream reaches of the River Kano', *Applied Geography* 4: 321-32.

O'Riordan, T. (1976a) *Environmentalism*, Pion, London (1st edition).

O'Riordan, T. (1976b) 'Beyond Environmental Impact Assessment', pp. 202-21 in T. O'Riordan and R. Hey (eds) *Environmental Impact Assessment*, Saxon House, Farnborough.

O'Riordan, T. (1981) *Environmentalism*, Pion, London (2nd. edition) .

O'Riordan, T. (1988) 'The politics of sustainability', pp. 29-50 in R.K. Turner (ed.) *Sustainable Environmental Management: principles and practice*. Westview Press, Boulder, Colo.

O'Riordan, T. and Hey, R. (eds) (1976) *Environmental Impact Assessment*. Saxon House, Farnborough. .

O'Riordan, T and Turner, R.K. (1983) *An Annotated Reader in Environmental Planning and Management*, Pergamon Press, Oxford.

Ormerod, W.E. (1978) 'The relationship between economic development and ecological degradation: how degradation has occurred in West Africa and how its progress might be altered', *Journal of Arid Environments* 1: 357-79.

Ormerod, W.E. (1986) 'A critical study of the policy of tsetse eradication', *Land Use Policy* 3: 85-99.

Osborn, F. (1948) *Our Plundered Planet*, Faber and Faber, London.

Osborn, F. (1954) *The Limits of the Earth*, Faber and Faber, London.

Osemiebo, G.J. (1988) 'Impacts of multiple forest land use on wildlife conservation in Bendel State, Nigeria', *Biological Conservation* 45: 209-20.

Osmaston, A.E. (1922) 'Notes on the forest communities of the Garhwal Himalaya', *Journal of Ecology* 10: 129-67.

Otten, M. (1986) 'Transmigrasi: from poverty to bare subsistence', *The Ecologist* 16: 57-67.

Otterman, J. (1974) 'Baring high-albedo soils by overgrazing: a hypothesised desertification mechanism', *Science* 166: 531-3.

Owens, S.E. (1986) 'Environmental politics in Britain: new paradigm or placebo?', *Area* 18: 195-201.

Page, W. and Richards, P. (1977) 'Agricultural pest control by community action: the case of the variegated grasshopper in southern Nigeria', *African Environment* 2/3: 127-41.

Palmer-Jones, R. (1984) 'Mismanaging the peasants: some origins of low productivity on irrigation schemes in northern Nigeria', in W.M. Adams and A.T. Grove (eds) *Irrigation in Tropical Africa: problems and problem-solving*, Cambridge African Monographs No. 3, University of Cambridge African Studies Centre, Cambridge.

Park, C.C. (1977) 'Man-induced changes in stream channel capacity', pp. 121-44 in K.J. Gregory (ed.) *River Channel Changes*, Wiley, Chichester.

Park, C.C. (1981) 'Man, river systems and environmental impact', *Progress in Physical Geography* 5: 1-31.

Park, C.C. (ed.) (1983) 'Environmental Impact Assessment', *ECOS: A Review of Conservation* 3(3): 1-41.

Park, C.C. (ed.) (1986) *Environmental Policies: an international review*, Croom Helm, London.

Parsons, H.L. (ed.) (1979) *Marx and Engels on Ecology*, Greenwood Press, Westport, Conn.

Passmore, J. (1980) *Man's Responsibility for Nature*, Duckworth, London.

Patten, C. (1987) 'Aid and the environment', *Geographical Journal* 153: 327-42.

Payne, A.J. (1986) *The Ecology of Tropical Lakes and Rivers*, Wiley, Chichester.

Pearce, D. (1988a) 'The sustainable use of natural resources in developing countries', pp. 102-17 in R.K. Turner (ed.) *Sustainable environmental management*, Belhaven, London.

Pearce, D. (1988b) 'Economists befriend the earth', *New Scientist* 19 November: 34-9.

Pearce, F. (1987) 'Pesticide deaths: the price of the Green Revolution', *New Scientist* 18 June: 30.

Pearl, R. (1927) 'The growth of populations', *Quarterly Review of Biology* 2: 537-43.

Pearsall, W.H. (1964) 'The development of ecology in Britain', *Journal of Ecology* 52: 1-12.

PEEM, (1987) 'Effects of agricultural development and change in agricultural practices on the transmission of vector-borne disease', pp. 8-51 in *Report of the 7th Meeting of the Joint WHO/FAO/UNEP Panel of Experts on Environmental Management for Vector Control (PEEM)*, PEEM Secretariat, Rome.

Penning-Rowsell, E.C. (1981) 'Consultancy and contract research', *Area* 13: 10-12.

Pepper, D. (1984) *The Roots of Modern Environmentalism*, Croom Helm, London.

Pepper, D. (1987) 'Environmental politics: who are the real radicals?', *Area* 19: 75-7.

Perfect, J. (1980) 'The environmental impacts of DDT in a tropical agro-ecosystem', *Ambio* 9: 16-22.

Pescod, M.B. (1978) 'Impact of development on the rural environment', in R.D. Hill and J.M. Bray (eds) *Geography and Environment in Southeast Asia*, Hong Kong University Press, Hong Kong.

Petr, T. (1975) 'On some factors associated with high fish catches in new African man-made lakes', *Archiv für Hydrobiologie* 75: 32-49.

Petr, T. (1979) 'Possible environmental impact on inland waters of two planned major engineering projects in Papua New Guinea', *Environmental Conservation* 6: 281-6.

Petts, G.E. (1977) 'Channel response to flow regulation: the case of the River Derwent, Derbyshire', pp. 145-64 in K.J. Gregory (ed.) *River Channel Changes*, Wiley, Chichester.

Petts, G.E. (1980) 'Implications of the fluvial process-channel morphology interaction below British reservoirs for stream habitats', *Science of the Total Environment* 16: 149-63.

Petts, G.E. (1984) *Impounded Rivers: perspectives for ecological management*, Wiley, Chichester.

Petulla, J.M. (1980) *American Environmentalism: values, tactics, priorities*, Texas A and M University Press, College Station, Tex. and London.

Phillips, J. (1931) 'Ecological investigations in South, Central and East Africa: outline of a progressive scheme', *Journal of Ecology* 14: 474-82.

Pickup, G. (1980) 'Hydrological and sediment modelling studies in the environmental impact assessment of a major tropical dam project', *Earth Surface Processes* 5: 61-75.

Pitt, D.C. (1976) *The Social Dynamics of Development*, Pergamon Press, Oxford.

Polunin, N. (1984) 'Genesis and progress of the World Campaign and Council for the Biosphere', *Environmental Conservation* 11: 293-8.

Poore, D. (1976) *Ecological Guidelines for Development in Tropical Rainforests*, IUCN, Gland, Switzerland.

Poore, D. and Sayer, J. (1987) *The Management of Tropical Moist Forest Lands: ecological guidelines*, IUCN, Gland, Switzerland.

Prince, S.D. (1986) 'Monitoring the vegetation of semi-arid tropical rangelands with NOAA-7 Advanced Very High Resolution Radiometer', pp. 307-34 in

M.J. Eden and J.T. Parry (eds) *Remote Sensing and Tropical Land Management*, Wiley, Chichester.
Prospero, J.M. and Nees, R.T. (1977) 'Dust concentrations in the atmosphere of the equatorial North Atlantic: possible relationship to the Sahelian drought', *Science* 196: 1196-8.
Prothero, R.M. (1962) 'Some observations on desiccation in north-western Nigeria', *Erkunde* 16: 112-19.
Pryde, P.R. (1972) *Conservation in the Soviet Union*, Cambridge University Press, Cambridge.
Rambo, A.T. (1982) 'Human ecology research on agroecosystems in SE Asia', *Singapore Journal of Tropical Geography* 3: 86-99.
Rapp, A. (1987) 'Desertification', pp. 425-43 in K.J. Gregory and D.E. Walling (eds) *Human Activity and Environmental Process*, Wiley, Chichester.
Rasid, H. (1979) 'The effects of regime regulation by the Gardiner Dam on downstream geomorphic processes in the South Saskatchewan River', *Canadian Geographer* 23: 140-58.
Ratcliffe, D.A. (1971) 'Criteria for the selection of nature reserves', *Advancement of Science* 27: 294-6.
Ravenhill, J. (1986) *Africa in Environmental Crisis*, Macmillan, London.
Raymond, H.L. (1968) 'Migration rates of yearling chinook salmon in relation to flows and impoundments in the Columbian and Snake rivers', *Transactions of the American Fisheries Society*.97: 3356-9.
Redclift, M. (1984) *Development and the Environmental Crisis: red or green alternatives?*, Methuen, London.
Redclift, M. (1987) *Sustainable Development: exploring the contradictions*, Methuen, London.
Redfern, M. (1986) 'Tsetse traps stop trypanosomiasis', *New Scientist* 27 February: 22.
Rees, W.A. (1978a) 'Do dams spell disaster for the Kafue Lechwe?', *Oryx* 14: 231-5.
Rees, W.A. (1978b) 'The ecology of the Kafue Lechwe as affected by the Kafue Gorge Hydroelectric scheme', *Journal of Applied Ecology* 15: 205-17.
Reich, C. (1970) *The Greening of America*, Random House, New York.
Reining, C.C. (1966) *The Zande Scheme: an anthropological case study of economic development in Africa*, Northwestern University Press, Evanston, Ill.
Repetto, R. (1987) 'Creating incentives for sustainable forest development', *Ambio* 16: 94-9.
Richards, P. (1985) *Indigenous Agricultural Revolution: ecology and food production in West Africa*, Longman, London.
Richards, P. (1986) *Coping with Hunger: hazard and experiment in an African rice-farming system*, Allen and Unwin, London.
Richards, P.W. (1939) 'Ecological studies on the rain forest of southern Nigeria; 1. The structure and floristic composition of the primary forest', *Journal of Ecology* 26: 1-61.
Riddell, R. (1981) *Ecodevelopment*, Gower, Aldershot.
Ridgeway, J. (1971) *The Politics of Ecology*, Dutton, New York.

Roberts, P. (1981) 'Rural development and the rural economy in Niger, 1900-1975', pp. 193-221 in J. Heyer, P. Roberts and G. Williams (eds) *Rural Development in Tropical Africa*, Macmillan, London.
Robinson, R.E. (1971) *Developing the Third World: the experience of the 1960s*, Cambridge University Press, Cambridge.
Roggeri, H. (1985) *African Dams: impacts on the environment*, Environment Liason Centre, Nairobi.
Roider, W. (1971) *Farm Settlements for Socio-economic Development: the Western Nigerian case*, Weltforum Verlag, Munich.
Rostow, W.W. (1978) *The World Economy: history and prospect*, Macmillan, London.
Roszak, T. (1979) *Person/planet: the creative disintegration of industrial society*, Victor Gollancz, London.
Routley, R. and Routley, V. (1980) 'Destructive forestry in Melanesia and Australia', *The Ecologist* 10: 56-7.
Rubin, N. and Warren, W.M. (eds) (1968) *Dams in Africa: an interdisciplinary study of man-made lakes in Africa*, Frank Cass, London.
Rudé, G. (1980) *Ideology and Popular Protest*, Lawrence and Wishart, London.
Rudig, W. and Lowe, P. (1986) 'The withered "greening" of British politics: a study of the Ecology Party', *Political Studies* 34: 262-84.
Rudig, W. and Lowe, P. (1989) *The Green Wave: a comparitive analysis of ecological parties*, Polity Press, London.
Russell, C.S. (1975) 'Environment and development', *Biological Conservation* 7: 227-37.
Russell, E.J. (1954) *World Population and World Food Supplies*, Allen and Unwin, London.
Rzóska, J. (ed.) (1976) *The Nile, Biology of an Ancient River*, Junk, The Hague.
Sachs, I. (1979) 'Ecodevelopment: a definition', *Ambio* 8 (2/3): 113.
Sachs, I. (1980) *Stratégies de l'écodéveloppement*, Les Editions Ouvrières, Paris.

Sagua, V.O. (1978) 'Flood control of the River Niger at Kainji Dam, Nigeria and its use during drought conditions, the period 1970-1976', pp. 127-40 in G. J. van Apeldoorn (ed.) *The Aftermath of the 1972-4 Drought in Nigeria*, Centre for Social and Economic Research, Ahmadu Bello University, Zaria, Nigeria.
Salisbury, E.J. (1964) 'The origin and early years of the British Ecological Society', *Journal of Ecology* 52: 13-18.
Sandbach, F. (1978) 'Ecology and the "limits to growth" debate', *Antipode* 10: 22-32.
Sandbach, F. (1980) *Environment, Ideology and Policy*, Basil Blackwell, Oxford.
Sandford, S. (1983) *Management of Pastoral Development in the Third World*, Wiley, Chichester.
Saulei, S.M. (1984) 'Natural regeneration following clear-fell logging operations in the Gogol Valley, Papua New Guinea', *Ambio* 13 (5-6): 351-4.
Schalie, H. van der (1960) 'Egypt's new high dam: asset or liability?', *Biologist* 43: 63-70.

Schlippe, P. de (1956) *Shifting Cultivation in Africa: the Zande system of agriculture*, Routledge and Kegan Paul, London.
Schmink, M. and Wood, C.H. (eds) (1984) *Frontier Expansion in Amazonia.*, University of Florida Press, Gainesville, Fla.
Schoepf, B.G. (1984) 'Man and the Biosphere in Zaire', pp. 269-90 in J. Barker (ed.) *The Politics of Agriculture in Tropical Africa*, Sage, Beverley Hills, Calif.
Schove, D.J. (1977) 'African droughts in the spectrum of time', pp. 38-53 in D. Dalby, R.J. Harrison Church and F. Bezzaz (eds) *Drought in Africa 2*, International African Institute, London.
Schumacher, E.F. (1973) *Small is Beautiful: economics as if people mattered*, Blond and Briggs, London (paperback edition published in 1974 by Abacus).
Scudder, T. (1975) 'Resettlement', pp. 453-71 in N.F. Stanley and M.P. Alpers (eds) *Man-made Lakes and Human Health*, Academic Press, London.
Scudder, T. (1980) 'River basin development and local initiative in African savanna environments', pp. 383-405 in D.R. Harris (ed.) *Human Ecology in Savanna Environments*, Academic Press, London,.
Scudder, T. (1988) *The African Experience with River Basin Development: achievements to date, the role of institutions and strategies for the future*, Institute of Development Anthropology and Clark University, Binghampton, NY.
Seabrook, A.T.P. (1957) 'The Groundnut Scheme in Retrospect', *Tanganyika Notes and Records* 47/48: 87-91.
Secretariat of the United Nations Conference on Desertification (eds) (1977) *Desertification: its causes and consequences*, Pergamon Press, Oxford.
Secrett, C. (1985a) *Rainforest: protecting the planet's richest resource*, Friends of the Earth, London.
Secrett, C. (1985b) 'Tropical rainforest', *Friends of the Earth Supporters' Newspaper*, June.
Seddon, D. (1983) *The Basic Needs Strategy in Theory and Practice: a utopian respose to world crisis?*, Development Studies Discussion Paper 119, University of East Anglia, Norwich.
Seddon, G. (1984) 'Logging in the Gogol Valley, Papua New Guinea', *Ambio* 13: 345-50.
Seeger, A. (1982) 'Native Americans and the conservation of flora and fauna in Brazil', pp. 177-90 in E.G. Hallsworth (ed.) *Socio-economic Effects and Constraints in Tropical Forest Management*, Wiley, Chichester.
Seers, D. (1977) 'The new meaning of development', *International Development Review* 3: 2-7.
Seers, D. (1980) 'North-South: muddling morality and mutuality', *Third World Quarterly* 2: 681-92.
Seeley, J. and Adams, W.M. (eds) (1987) *Environmental Issues in African Development Planning*, Cambridge African Monograph No. 9, University of Cambridge African Studies Centre, Cambridge.
Shaikh, R.A. and Nichols, J.K. (1984) 'The international management of chemicals', *Ambio* 13: 88-92.
Shalash, S. (1983) 'Degradation of the River Nile', Parts 1 and 2, *Water Power and Dam Construction* 35(7): 7-43 and 35(8): 56-8.

Shantz, H.L. and Marbut, C.F. (1923) *The Vegetation and Soils of Africa*, American Geographical Society Research Series 13, American Geographical Society and National Research Council, New York.

Sharaf el Din, S.H. (1977) 'Effects of the Aswan High Dam on the Nile flood on the estuarine and coastal circulation pattern along the Mediterranean Egyptian coast', *Limnology and Oceanography* 22: 194-207.

Shaw, P. (1985) 'Late Quaternary landforms and environmental change in northwest Botswana: the evidence of Lake Ngami and the Mababe Depression', *Transactions of the Institute of British Geographers N.S.* 10: 333-46.

Sheail, J. (1975) 'The concept of National Parks in Great Britain 1900-1950', *Transactions of the Institute of British Geographers* 66: 41-56 .

Sheail, J. (1976) *Nature in Trust: the history of nature conservation in Great Britain*, Blackie, Glasgow.

Sheail, J. (1981) *Rural Conservation in Inter-War Britain*, Clarendon Press, Oxford.

Sheail, J. (1984a) 'Nature reserves, national parks and post-war reconstruction in Britain', *Environmental Conservation* 11: 29-34.

Sheail, J. (1984b) 'Constraints on water resource development in England and Wales: the concept and management of compensation flows', *Journal of Environmental Management* 19: 351-62.

Sheail, J. (1985) *Pesticides and Nature Conservation: the British experience 1950-1975*, Clarendon Press, Oxford.

Sheail, J. (1987) *Seventy-five Years of Ecology: the British Ecological Society* . Blackwell Scientific, Oxford.

Shelton, N. (1985) 'Logging versus the natural habitat in the survival of tropical forests', *Ambio* 14 (1): 39-42.

Shenton, R. (1986) *The Development of Capitalism in Northern Nigeria*, James Currey, London.

Sheppe, W.A. (1985) 'Effects of human activities on Zambia's Kafue Flats ecosystem', *Environmental Conservation* 12: 49-57.

Shiva, V. (1988) *Staying Alive: women, ecology and development*, Zed Books, London.

Silberfein, M. (1978) 'The African cultivator: a geographic overview', pp. 7-23 in A.K. Smith and C.E. Welch (eds) *Peasants in Africa*, Crossroads Press, Waltham, Mass. for the African Studies Association.

Simon, J.L. (1981) *The Ultimate Resource*, Princeton University Press, Princeton, NJ.

Simon, J.L. and Kahn H. (eds) (1984) *The Resourceful Earth: a response to Global 2000*, Blackwell, Oxford.

Simons, D.B. and Li, R.-M. (1982) 'Bank erosion on regulated rivers', pp. 717-54 in R.D. Hey, J.C. Bathurst, and C.R. Thorne (eds) *Gravel Bed Rivers*, Wiley, Chichester.

Sinclair, A.R. and Fryxell, J.M. (1985) 'The Sahel of Africa: ecology of a disaster', *Canadian Journal of Zoology* 63: 987-94.

Singh, A. (1986) 'Change detection in the tropical forest environment of northeast India using Landsat', pp. 237-54 in M.J. Eden and J.T. Parry (eds) *Remote Sensing and Tropical Land Management*, Wiley, Chichester.

Singh, G. (1980) 'Ecodevelopment: the environmental movement's viewpoint, pp. 1349-55 in J. Furtado (ed.) *Tropical Ecology and Development*, International Society for Tropical Ecology, Kualar Lumpur.

Singh, J.S., Singh, S.P., Saxena, A.K. and Rawat, Y.S. (1984) 'India's Silent Valley and its threatened rainforest ecosystems', *Environmental Conservation* 11: 223-33.

Sioli, H. (1985) 'The effects of deforestation in Amazonia', *Geographical Journal* 151: 197-203.

Skillings, R.F. (1984) 'Economic development of the Brazilian Amazon: opportunities and constraints', *Geographical Journal* 150: 48-54.

Smil, V. (1984) *The Bad Earth: environmental degradation in China*, Sharp, Armouh, NY.

Smith, A.K. and Welch, C.E. (eds) (1978) *Peasants in Africa*, Crossroads Press, Waltham, Mass. for the African Studies Association.

Smith, N. (1984) *Uneven Development*, Basil Blackwell, Oxford.

Soerianegara, I. (1982) 'Socio-economic aspects of forest resources management in Indonesia', pp. 73-85 in E.G. Hallsworth (ed.) *Socio-economic Effects and Constraints in Tropical Forest Management*, Wiley, Chichester.

Speth, J.B. (1984) 'The biospheral possible', *Environmental Conservation* 11: 351-4.

Spittler, G. (1979) 'Peasants and the state in Niger (West Africa)', *Journal of Peasant Studies* 7: 30-47.

Stamp, L.D. (1925) 'The aerial survey of the Irrawaddy Delta forests (Burma)', *Journal of Ecology* 13: 262-76.

Stamp, L.D. (1938) 'Land utilisation and soil erosion in Nigeria', *Geographical Review* 28: 32-45.

Stamp, L.D. (1940) 'The southern margin of the Sahara: comments on some recent studies on the question of desiccation', *Geograpical Review* 30: 297-300.

Stamp, L.D. (1953) *Our Undeveloped World*, Faber and Faber, London.

Stebbing, E.P. (1935) 'The encroaching Sahara: the threat to the West African Colonies', *Geographical Journal* 85: 506-24.

Stebbing, E.P. (1938) 'The man-made desert of Africa: erosion and drought', *Journal of the Royal African Society Supplement*, January.

Stein, R.E. and Johnson, B. (1979) *Banking on the Biosphere? Environmental procedures and practices of nine multilateral aid agencies*, Lexington Press, New York.

Stewart, F. (1975) 'A note on social cost-benefit analysis and class conflict in LDCs', *World Development* 3 (1): 31-39.

Stewart, F. (1985) *Planning to Meet Basic Needs*, Macmillan, London.

Stock, R.F. (1978) 'The impact of the Hadejia River floods in Hadejia Emirate, Kano State', pp. 141-6 in G. Jan van Apeldoorn (ed.) *The Aftermath of the 1972-74 Drought in Nigeria: Proceedings of a conference held at Bagauda, April 1977*, Federal Department of Water Resources and the Centre for Social and Economic Research, Ahmado Bello University, Zaria, Nigeria.

Stoddart, D.R. (1970) 'Our environment', *Area* 2(1): 1-3.

Stoddart, D.R. (1986) *On Geography and its History*, Blackwell, Oxford .

Stohr, W.B. (1981) 'Development from below: the bottom-up and periphery-inward development paradigm', pp. 39-72 in W.B. Stohr and D.R.F. Taylor (eds) *Development: from above or below?*, Wiley, Chichester.

Stolper, W. (1966) *Planning Without Facts*, Harvard University Press, Cambridge, Mass.

Stone, J.C. (ed.) (1980) *Experts in Africa*, University of Aberdeen African Studies Group, Aberdeen.

Strange, S. (1986) *Casino Capitalism*, Blackwell, Oxford.

Street, F.A. (1981) 'Tropical palaeoenvironments', *Progress in Physical Geography* 5: 157-85.

Street, F.A. and Grove, A.T. (1976) 'Environmental and climatic implications of late Quaternary lake-level fluctuations in Africa', *Nature* 261: 385-90.

Street-Perrott, F.A., Roberts, N. and Metcalfe, S. (1985) 'Geomorphic implications of late Quaternary hydrological and climatic changes in the northern Hemisphere tropics', pp. 165-83 in I. Douglas and T. Spencer (eds) *Environmental Change and Tropical Geomorphology*, Allen and Unwin, Hemel Hempstead.

Streeten, P., Burki, J., ul Haq, M., Hicks, N. and Stewart, F. (1981) *First Things First: meeting basic human needs in developing countries*, World Bank and Oxford University Press, London.

Stutley, P.W. (1980) 'Some aspects of the contemporary scene', pp. 76-95 in J. C. Stone (ed.) *Experts in Africa*, University of Aberdeen African Studies Group, Aberdeen.

Suckcharoen, S., Nuorteva, P. and Hasanen, E. (1978) 'Alarming signs of mercury pollution in a freshwater area of Thailand', *Ambio* 7: 113-16.

Sugden, R. and Williams, A. (1978) *The Principles of Practical Cost-benefit Analysis*, Oxford University Press, Oxford.

Sulton, I. (1970) 'Dams and the environment', *Geographical Review* 60: 128- 9.

Survival International (1985) *An End to Laughter? Tribal peoples and economic development*, Review 44, Survival International, London.

Sutcliffe, R.B. (1972) *Industry and Underdevelopment*, Addison Wesley, London.

Sutcliffe, R.B. (1984) 'Industry and underdevelopment re-examined', *Journal of Development Studies* 21: 121-33.

Sutlive, V.H., Altshuler, N. and Zamora, M.D. (eds) (1980) *Where Have All The Flowers gone? Deforestation in the Third World*, Studies in Third World Societies No. 13, College of William and Mary, Williamsburg, Va.

Sutlive, V.H., Altshuler, N. and Zamora, M.D. (eds) (1981) *Blowing In The Wind: deforestation and long-range implications*, Studies in Third World Societies No. 14, College of William and Mary, Williamsburg, Va.

Sutton, J. (1984) 'Irrigation and soil conservation in African agricultural history: with a reconsideration of the Inyanga terracing (Zimbabwe) and Engaruka irrigation works (Tanzania)', *Journal of African History* 25: 25-41.

Sutton, J. (1986) 'The irrigation and manuring of the Engaruka field system: further observations and historic discussion of a later Iron Age settlement settlement in the Northern Tanzanian Rift', *Azania* 21: 27-51.

Sutton, K. (1977) 'Population resettlement - traumatic upheavals and the Algerian experience', *Journal of Modern African Studies*, 15: 279-300.

Svedin, U. (1987) 'The IUCN Conference on conservation and development in Ottawa', *Ambio* 16(1): 65.

Swales, S. and O'Hara, K. (1980) 'Instream habitat improvement devices and their use in freshwater fisheries management', *Journal of Environmental Management* 10: 167-179.

SWECO (1978) *Great Ruaha Power Project, Tanzania: expected environmental changes below the Mtera reservoir*, SWECO, Stockholm.

Tahir, A.A. (1980) 'The Sudd as a wetland ecosystem and the Jonglei Canal Project', *Water Supply and Management* 4: 53-4.

Talbot, M.R. (1980) 'Environmental responses to climatic change in the West African Sahel over the past 20,000 years', pp. 37-42 in M.A.J. Williams and H. Faure (eds) *The Sahara and the Nile: Quaternary environments and prehistoric occupation in northern Africa*, Balkema, Rotterdam.

Talling, J.F. (1980) 'Some problems of aquatic environments in Egypt from a general viewpoint of Nile ecology', *Water Supply and Management* 4: 13-20.

Tamakloe, M.A. (1968) 'New Mpamu: case study of a Volta River Authority resettlement town', *Ghana Journal of Agricultural Science* 1: 45-51.

Taylor, T.A. (1977) 'Mixed cropping as an input in the management of crop pests in tropical Africa', *African Environment* 2/3: 111-26.

Thangan, E.S. (1982) 'Problems of shifting cultivation in North-eastern India', pp. 53-63 in E.G. Hallsworth (ed.) *Socio-economic Effects and Constraints in Tropical Forest Management*, Wiley, Chichester.

Thomas, D.S.G. (1984) 'Ancient ergs of the former arid zones of Zimbabwe, Zambia and Angola', *Transactions of the Institute of British Geographers N.S.* 9: 75-88.

Thomas, K. (1983) *Man and the Natural World: changing attitudies in England 1500-1800*, Allen Lane, London (paperback Penguin, Harmondsworth 1984).

Thomas, W.L. (ed.) (1956) *Man's Role in Changing the Face of the Earth*, University of Chicago Press, Chicago.

Thompson, W.S. (1956) 'The spiral of population', pp. 970-86 in W.L. Thomas *Man's Role in Changing the Face of the Earth*, University of Chicago Press, Chicago.

Throup, D.W. (1987) *Economic and Social Origins of Mau Mau*, James Currey, London.

Tignor, R.L. (1971) 'Kamba political protest: the destocking controversy of 1938', *International Journal of African Historic Studies* 4: 237-51.

Timberlake, L. (1985) *Africa in Crisis: the causes, the cures of environmental bankruptcy*, Earthscan, London.

Timberlake, L. (1987) *Only One Earth: living for the future*, BBC and Earthscan, London.

Tolba, M.K. (1986) 'Desertification in Africa', *Land Use Policy* 3: 260-8.

Trenbath, B.R. (1984) 'Decline of soil fertility and the collapse of shifting cultivation systems under intensification', pp. 279-92 in A.C. Chadwick and S.C. Sutton (eds) *Tropical Rain-forest: the Leeds symposium*, Leeds Philosophical Society, Leeds.

Tricart, J. (1985) 'Evidence of upper Pleistocene dry climates in northern South America', pp. 197-217 in I. Douglas and T. Spencer (eds) *Environmental Change and Tropical Geomorphology*, Allen and Unwin, Hemel Hempstead.

Trislsbach, A. and Hulme, M. (1984) 'Recent rainfall changes in central Sudan and their physical and human implications', *Transactions of the Institute of British Geographers N.S.* 9: 280-98.

Tucker, C.J., Holben, B.N. and Goff, T.E. (1984) 'Intensive forest clearing in Rondonia, Brazil, as detected by satellite remote sensing', *Remote Sensing of the Environment* 15: 255-61.

Tucker, C.J., Townshend, J.E.G. and Goff, T.E. (1985) 'African land cover classification using satellite data', *Science* 277: 369-75.

Tucker, C.J., Justice, C.O. and Prince, S.D. (1986) 'Monitoring the grasslands of the Sahel 1984-1985', *International Journal of Remote Sensing* 7: 1571-83.

Tuntamiroon, N. (1985) 'The environmental impact of industrialisation', *The Ecologist* 15(4): 161-4.

Turner, B.L. II and Brush, S.B. (1987) (eds) *Comparitive Farming Systems*, Guildford Press, Hove.

Turner, D. J. (1971) 'Dams in ecology', *Civil Engineering* 41: 76-80.

Turner, R.K. (1988a) (ed.) *Sustainable Environmental Management: principles and practice*, Westview Press, Boulder, Colo.

Turner, R.K. (1988b) 'Sustainability, resource conservation and pollution control: an overview', pp. 17-25 in R.K. Turner (ed.) *Sustainable Environmental Management: principles and practice*, Westview Press, Boulder, Colo.

Turton, D. (1987) 'The Mursi and National Park development in the lower Omo Valley', pp. 169-86 in D.M. Anderson and R.H. Grove (eds) *Conservation in Africa: people, policies and practice*, Cambridge University Press, Cambridge.

Tyson, P.D., Dyer, T.G.J. and Mametse, M.N. (1975) 'Secular changes in South African rainfall, 1880-1972', *Quarterly Journal of the Royal Meteorological Society* 101: 817-33.

UK Programme Organising Committee (1983) *The Conservation and Development Programme for the UK: a response to the World Conservation Strategy*, Kogan Page, London.

Underhill, H.W. (1984a) *Small Scale Irrigation in Africa in the Context of Rural Development*, Food and Agriculture Organisation, Rome.

Underhill, H.W. (1984b) 'The roles of governmental and non-governmental organisations and international agencies in small holder irrigation development', pp. 9-23 in M.J. Blackie (ed.) *African Regional Symposium on Small Holder Irrigation*, Hydraulics Research Ltd, Wallingford.

UNEP (1978) *Review of Areas: environment and development and environmental management*, Report No. 3, United Nations Environment Programme, Nairobi.

UNESCO (1963) *A Review of the Natural Resources of the African Continent*, UNESCO, Paris.

UNESCO (1973) *Programme on Man and the Biosphere (MAB). Expert Panel on Project 8: conservation of natural areas and of the genetic material they contain, Final Report*, MAB Report 12, UNESCO, Paris.

UNESCO (1976a) *Programme on Man and the Biosphere (MAB). Expert Panel on Project 4: impact of human activities on the dynamics of arid and semi-*

arid zone ecosystems, with particular attention to the effects of irrigation, Final Report, MAB Report 29, UNESCO, Paris.

UNESCO (1976b) *Programme on Man and the Biosphere (MAB). Expert Consultation on Project 10: effects on man and his environment of major engineering works, Final Report*, Man and the Bisophere Report Series No. 37, UNESCO, Paris.

UNESCO (not dated) 'Man and the Biosphere Programme', pamphlet, UNESCO, Paris

United Nations (ed.) (1977) *Desertification: its causes and consequences*, Pergamon Press, Oxford.

Vaidyanathan, A. (1985) 'A sorry technological tale from India', *New Scientist* 21 February.

Verstraete, M.M. (1986) 'Defining desertification: a review', *Climatic Change* 9: 5-18.

Vickers, W.T. (1984) 'Indian policy in Amazonian Ecuador', pp. 8-32 in M. Schmink and C.H. Wood (eds) *Frontier Expansion in Amazonia*, University of Florida Press, Gainesville, Fla.

Vogt, W. (1949) *Road to Survival*, Victor Gollancz, London.

Waddington, C.H. (1975) 'The origin', pp. 4-11 in E.B. Worthington (ed.) *The Evolution of the IBP*, Cambridge University Press, Cambridge.

Wade, R. (1982) 'The system of administrative and political corruption: canal irrigation in south India', *Journal of Development Studies* 18: 287-328.

Wallace, T. (1980) 'Agricultural projects and land in Northern Nigeria', *Review of African Political Economy* 17: 59-70.

Wallace, T. (1981) 'The Kano River Project: the impact of an irrigation scheme on productivity and welfare', in J. Heyer, P. Roberts, and G. Williams, (eds) *Rural Development in Tropical Africa*, Macmillan, London.

Wallerstein, I. (1974) *The Modern World System 1: Capitalist agriculture and the origins of the European world-economy in the sixteenth century*, Academic Press, London.

Walls, J. (1984) 'Summons to action', *Desertification Control Bulletin* 10: 5-14.

Walsh, J. (1985) 'Onchocerciasis: river blindness', pp. 269-94 in A.T. Grove (ed.) *The Niger and its Neighbours: environment, history and hydrobiology, human use and health hazards of the major West African rivers*, Balkema, Rotterdam.

Walter, I. and Ugelow, J.L. (1979) 'Environmental problems in developing countries', *Ambio* 8: 102-9.

Ward, B. and Dubos, R. (1972) *Only One Earth: the care and maintenance of a small planet*, André Deutsch, London.

Ward, J.V and Stanford, J.A. (eds) (1979) *The Ecology of Regulated Streams*, Plenum Press, New York.

Warren, A. and Agnew, C. (1988) *An Assessment of Desertificaton and Land Degradation in Arid and Semi-arid Areas*, Drylands Programme Paper No. 2, International Institute for Environment and Development, London.

Warren, A. and Maizels, J.K. (1977) 'Ecological change and desertification', pp. 169-261 in Secretariat of the United Nations Conference (eds) *Desertification: its causes and consequences*, Pergamon Press, Oxford.

Warren, B. (1980) *Imperialism: pioneer of capitalism*, New Left Books/ Verso, London.

Washbourn, C. (1967) 'Lake levels and Quaternary climates in the eastern Rift Valley of Kenya', *Nature* 216: 672-3

Waterbury, J. (1979) *Hydropolitics of the Nile Valley*, Syracuse University Press, New York.

Wathern P. (ed.) (1988) *Environmental Impact Assessment: theory and practice*, Unwyn Hyman, London.

Watson, C. (1986) 'Working at the World Bank', pp. 268-75 in T. Hayter and C. Watson (eds) *Aid: rhetoric and reality*, Pluto Press, London.

Watts, M.J. (1983a) *Silent Violence: food, famine and peasantry in northern Nigeria*, University of California Press, Berkeley, Calif.

Watts, M.J. (1983b) 'On the poverty of theory: natural hazards research in context', pp. 231-62 in K. Hewitt (ed.) *Interpretations of Calamity: from the viewpoint of human ecology*, Allen and Unwin, London.

Watts, M J. (1984) 'The demise of the moral economy: food and famine in the Sudano-Sahelian region in historical perspective', pp. 128-48 in E.P. Scott (ed.) *Life Before the Drought*, Allen and Unwin, London.

Watts, M.J. (1987) 'Drought, environment and food security: some reflections on peasants, pastoralists and commoditisation in dryland East Africa', pp. 171-217 in M.H. Glantz (ed.) *Drought and Hunger in Africa: denying famine a future*, Cambridge University Press, Cambridge.

Welcomme, R.L. (1979) *The Fisheries Ecology of Floodplain Rivers*, Longman, London.

Wellings, P. (1983) 'Making a fast buck in Lesotho: capital leakage and the public accounts of Lesotho', *African Affairs* 82: 495-507.

Wendler, G. and Eaton, F. (1983) 'On the desertification of the Sahel Zone; Part 1: ground observations', *Climatic Change* 5: 365-80.

Western, D. (1982) 'Amboseli National Park: enlisting landowners to conserve migratory wildlife', *Ambio* 11: 302-8.

Westing, A.H. (1981) 'A world in balance', *Environmental Conservation* 8: 177-83.

White, E. (1969) The place of biological research in the development of the resources of man-made lakes', pp. 37-48 in E.L. Obeng (ed.) *International Symposium on Man-made Lakes, Accra*, Ghana Universities Press, Accra.

White, S. (1978) 'Cedar and mahogany logging in eastern Peru', *Geographical. Review* 68: 394-416.

Whitton, B.A. (1975) *River Ecology*, Blackwell, London.

Williams, G. (1976) 'Taking the part of the peasants: rural development in Nigeria and Tanzania', pp. 131-54 in P. Gutkind and I. Wallerstein (eds) *The Political Economy of Contemporary Africa*, Sage, Beverely Hills.

Williams, G. (1981) 'The World Bank and the peasant problem', pp. 16-51 in J. Heyer, P. Roberts, and G. Williams (eds) *Rural Development in Tropical Africa*, Macmillan, London.

Williams, G.J. (1977) 'The Kafue Hydroelectric Scheme and its environmental setting', pp. 13-27 in G.J. Williams and G.W. Howard (eds) *Development and Ecology in the Lower Kafue Basin in the Nineteen Seventies*, Kafue Basin Research Committee, Lusaka.

Williams, G.J. and Howard, G.W. (eds) (1977) *Development and Ecology in the Lower Kafue Basin in the Nineteen Seventies*, Kafue Basin Research Committee, Lusaka. .
Williams, G.P. and Wolman, M.G. (1984) *Downstream Effects of Dams on Alluvial Rivers*, USGS Professional Paper 1286, United States Geological Survey, Washington, DC.
Williams, M.A.J. (1985) 'Pleistocene aridity in tropical Africa, Australia and Asia', pp. 219-38 in I. Douglas and T. Spencer (eds) *Environmental Change and Tropical Geomorphology*, Allen and Unwin, Hemel Hempstead.
Winid, B. (1981) 'Comments on the development of the Awash Valley, Ethiopia', pp. 147-65 in S.K. Saha and C.J. Barrow (eds) *River Basin Planning: theory and practice*, Wiley, Chichester.
Winstanley, D. (1973a) 'Recent rainfall trends in Africa, the Middle East and India', *Nature* 243: 464-5.
Winstanley, D. (1973b) 'Rainfall patterns and general atmospheric circulation', *Nature* 245: 190-4.
Winterbottam, R. and Hazelwood, P.T. (1987) 'Agroforestry and sustainable development: making the connection', *Ambio* 16 (2-3): 100-10.
Wood, A. (1950) *The Groundnut Affair*, Bodley Head, London.
Wood, C. (1982) 'Implications of the European EIA initiative', *ECOS: A Review of Conservation* 3(3): 25-9.
World Bank (1981) *Accelerated Development in Sub-Saharan Africa*, World Bank, Washington, DC.
World Bank (1984a) *Tribal Peoples and Economic Development: human ecologic considerations*, World Bank, Washington, DC.
World Bank (1984b) *World Development Report 1984*, Oxford University Press, London and New York.
World Bank (1984c) *Environmental Policies and Procedures of the World Bank*, Office of Environmental and Health Affairs, World Bank, Washington, DC.
World Resources Institute and IIED (1986) *World Resources 1986*, Basic Books, New York.
World Water (1979) 'Mekong: power not politics', *World Water* November: 20-5.
World Wild Fund for Nature (1988) 'Debt-for-nature swap in the Philippines', *WWF News* July/August: 6.
Worster, D. (1979) *Dust Bowl: the southern plains in the 1930s*, Oxford University Press, Oxford.
Worster, D. (1985) *Nature's Economy: a history of ecological ideas*, Cambridge University Press, Cambridge.
Worthington, E.B. (1932) *A Report on the Fisheries of Uganda Investigated by the Cambridge Expedition to the East African Lakes 1932-33*, Crown Agents for the Colonies, London.
Worthington, E.B. (1938) *Science in Africa: a review of scientific research relating to tropical and southern Africa*, Royal Institute of International Affairs, London.
Worthington, E.B. (1958) *Science in the Development of Africa: a review of the contribution of physical and biological knowledge south of the Sahara*,

Commission for Technical Cooperation in Africa South of the Sahara and the Scientific Council for Africa South of the Sahara, London.

Worthington, E.B. (ed.) (1975) *The Evolution of the IBP*, Cambridge University Press, Cambridge.

Worthington, E.B. (1982a) 'World Campaign for the Biosphere', *Environmental Conservation* 9: 93-100.

Worthington, E.B. (1982b) 'The twentieth anniversary of the IBP', Environmental Conservation 9: 70.

Worthington, E.B. (1983) *The Ecological Century: a personal appraisal*, Cambridge University Press, Cambridge.

Young, A.M. (1986) 'Eco-enterprises: eco-tourism and farming of exotics in the tropics', *Ambio* 15: 361-3.

Yuqian, L. and Qishun, Z. (1981) 'Sediment regulation problems in Sanmenxia Reservoir', *Water Supply and Management* 5: 351-60.

Zhao, D. and Sun, B. (1986) 'Air pollution and acid rain in China', *Ambio* 15(1): 2-5.

Index